LUTON SIXTH FOR

LUTO

KU-285-837

Illustrated Biology

B. S. Beckett

1478
1978

Oxford University Press

179423

Oxford University Press, Walton Street, Oxford OX2 6DP

Oxford London Glasgow
New York Toronto Melbourne Wellington
Kuala Lumpur Singapore Jakarta Hong Kong Tokyo
Delhi Bombay Calcutta Madras Karachi
Ibadan Nairobi Dar es Salaam Cape Town

ISBN 0 19 914044 8

© Oxford University Press 1978

First published 1978
Reprinted 1978 and 1979

B. S. Beckett

The author is a graduate of Hull University where he
obtained a B.Sc. degree in Zoology and Botany in
1963. After that he taught in schools in Leeds and
Hull, and from 1968–71 he was Head of Biology at
King George V School in Hong Kong. While he was
at this last post he carried out investigations into the
use and evaluation of discovery methods in science
teaching, which led to an M.A. (Ed.) degree. When he
returned to England he continued his research on
discovery and formal teaching methods, for which he
was awarded a B.Phil. degree by Hull University.

   Since then the author has divided his time between
writing, illustrating, and teaching.

Filmset in 'Monophoto' Plantin 110 11 on 12 pt by
Tradespools Limited, Frome, Somerset
and printed in Great Britain by
Lowe and Brydone Printers Ltd., Thetford, Norfolk

Contents

Acknowledgements

The publishers would like to thank the following for permission to reproduce photographs:

Animal Photography Ltd., p. 153, top right, second from top, second from bottom; Ardea London, pp. 119 (Dr. K. J. Carlson), 176 (P. Morris); British Museum (Natural History), p. 150; Camera Press Ltd., p. 176, top right; Dr. K. E. Carr and Dr. P. G. Toner, p. 49; Bruce Coleman, pp. 121, 154, top left; Fox Photos Ltd., p. 176, top right; B. J. Harris, p. 140;

Keystone Press Agency, p. 71; NHPA, pp. 123 (L. E. Perkins), 153, bottom right; Pictorial Press Ltd., p. 120; Picturepoint Ltd., pp. 154, middle left, 176, top left, 177, top right; Pigeons and Pigeon World, p. 154, bottom right, top right, bottom left; RSPB, p. 177, bottom right; J. R. Tabberner, p. 140.

The diagram on p. 111 has been adapted from *How Life Begins* by Reid and Booth, 1971, published by Heinemann Educational Books Limited.

Preface

This book is designed specifically for CSE students and covers the main topics required by all CSE syllabuses. It will also serve as a useful supplementary or 'foundation' text for 'O' level candidates.

The book consists of 90 double-page units, each of which presents a self-contained explanation of one particular topic. The left-hand page of each of these units contains about 400 words of text; the remaining space (usually about a page and a half) is devoted to drawings, diagrams, or photographs.

The book contains two types of illustration, which are most clearly contrasted in the pages dealing with the structure of organisms. In these units the right-hand page contains realistic drawings designed to attract attention, generate interest, and give a clear impression of specimens as they appear in nature. The left-hand pages contain simplified versions of the realistic drawings, or a diagrammatic summary of the biological principles covered by the unit.

The text is written in simple language. Technical terms are fully explained and printed in bold type when they first appear, so that they can easily be revised in context. In addition a glossary is provided to help students learn technical terms.

Units covering general topics such as classification, transport, co-ordination, and reproduction are grouped together. This approach, which is followed in modern integrated syllabuses, makes it easy to compare biological characteristics in a variety of different organisms.

At the end of the book there is a section of revision tests. The tests are of two kinds: vocabulary tests enable students to check their knowledge of technical terms used in the text, and comprehension tests establish whether or not they have understood and assimilated what they have read.

1

Protists
(simple organisms)

Protozoa

Protozoa are microscopic unicellular organisms. The main types are:

Rhizopods Protozoa which move and usually feed by means of pseudopodia. Some are enclosed in a shell of lime (e.g. *Foraminiferans*). Others have no shell (e.g. *Amoeba*).

Ciliates Protozoa with microscopic hairs called cilia, which are used in feeding and movement (e.g. *Vorticella, Paramecium*).

Flagellates Protozoa which move by means of whiplike flagella. Many are green (e.g. *Euglena*) and live by photosynthesis.

Fungi

Most fungi are multicellular, although yeast is an important exception, since it is unicellular. Multicellular fungi are usually made up of fine threads called hyphae. These hyphae are collectively known as the mycelium of the fungus. Many fungi are saprophytes (e.g. *Mucor*, mushrooms). This means they feed by decomposing food and dead organisms. Other fungi are parasites, causing diseases such as mildew in plants, and ringworm in animals.

Algae

Algae are simple plant-like organisms. They live by photosynthesis. Green algae live in the sea, in fresh water, or in damp places on land. They occur in various forms: as single cells, hollow balls of cells, fine threads (e.g. *Spirogyra*), or hollow tubes. Brown algae live in the sea. They often reach several metres in length (e.g. bladderwrack, kelp). Diatoms are unicellular algae. Their cell walls are made of silica, and consist of two pieces which fit together like a box and lid.

Protozoa   Amoeba × 70

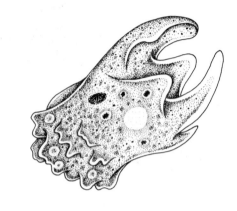

Fungi   Mucor (pin mould) × 65

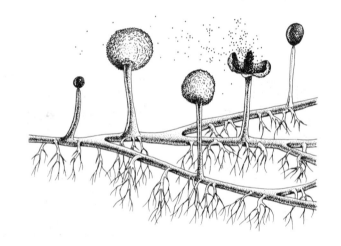

Algae   Spirogyra × 80

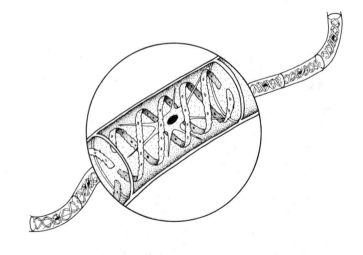

2

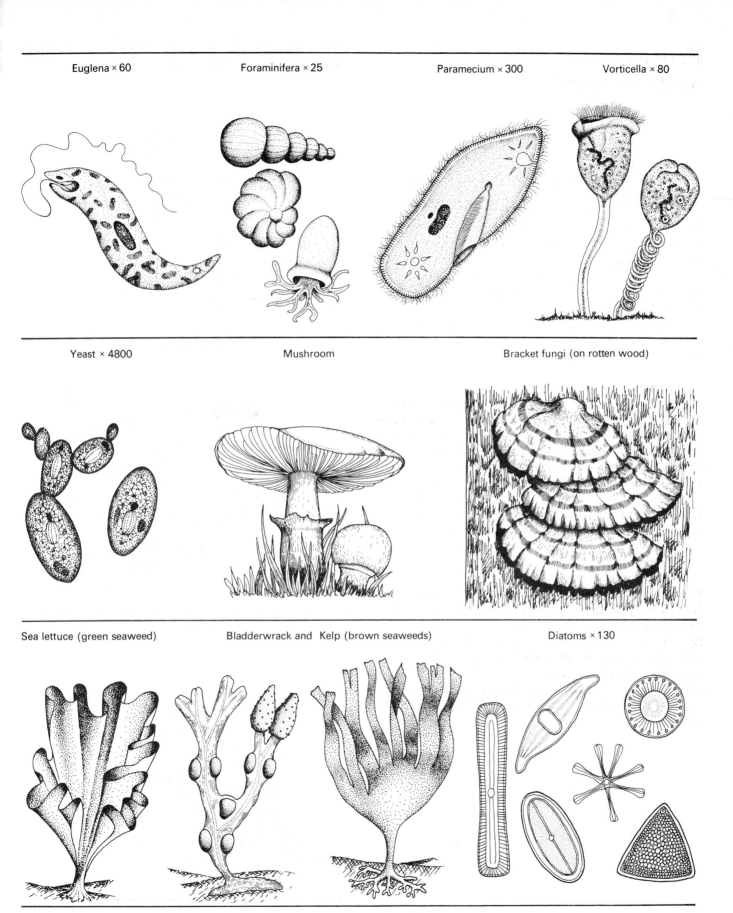

Euglena × 60   Foraminifera × 25   Paramecium × 300   Vorticella × 80

Yeast × 4800   Mushroom   Bracket fungi (on rotten wood)

Sea lettuce (green seaweed)   Bladderwrack and Kelp (brown seaweeds)   Diatoms × 130

3

Invertebrates
(animals without backbones)

Coelenterates

Coelenterates live in water. Sea anemones, jelly-fish, and corals are examples found in the sea, and *Hydra* lives in ponds and streams. Coelenterates have a hollow tubular body with one opening. This is the mouth, which is usually surrounded by tentacles. The body and tentacles of coelenterates are covered with sting cells. These are used to paralyse the prey, which is then pulled into the mouth by the tentacles.

Platyhelminths (flatworms)

Flatworms have flat bodies with a single opening, the mouth. Unlike true worms (e.g. earthworms), flatworms do not consist of segments. *Planarians* are flatworms which live in ponds, streams, and damp soil. Tapeworms and liver flukes are parasitic flatworms. They live in man and other animals.

Annelids (true worms)

Annelid worms have bodies made up of many similar segments. The boundaries of the segments are marked by external grooves around the body. Some live in the soil (e.g. earthworms), some in the sea under sand or mud (e.g. sandworms), and others are parasites (e.g. leeches).

Coelenterates

Hydra × 3

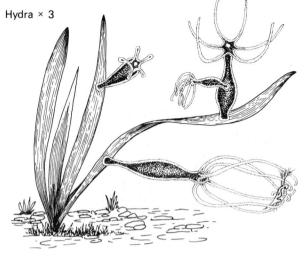

Platyhelminths (flatworms)

Planarian (freshwater flatworm)

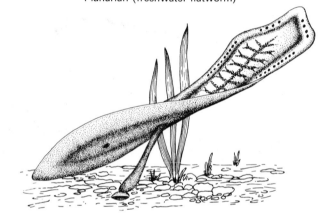

Annelids (true worms)

Earthworm

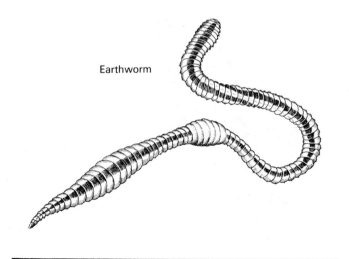

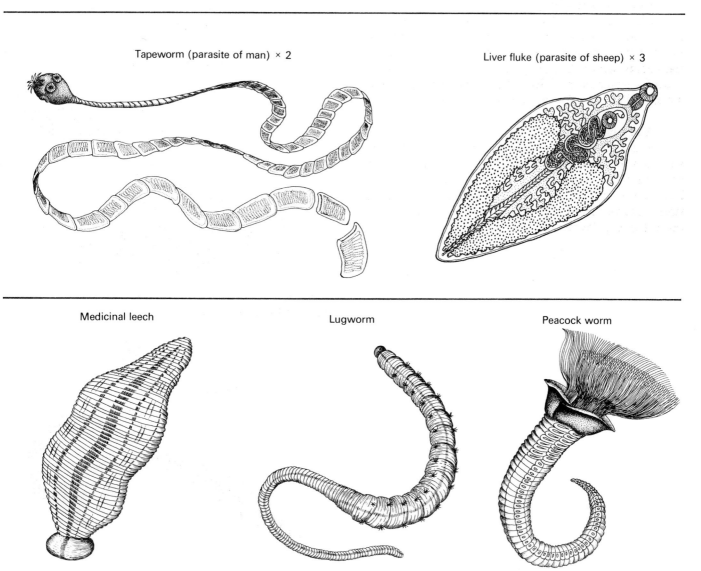

Jellyfish

Sea anemones (closed and open)

Tapeworm (parasite of man) × 2

Liver fluke (parasite of sheep) × 3

Medicinal leech

Lugworm

Peacock worm

3

Invertebrates (*continued*)

Arthropods

Arthropods have a segmented body enclosed in a jointed exoskeleton. They all have several pairs of legs, and many have compound eyes and antennae. The main types are:

Crustaceans Arthropods with two pairs of antennae. Many have a hard chalky exoskeleton (e.g. crab, lobster); others have a thinner exoskeleton (e.g. woodlouse, flea).

Arachnids Arthropods with four pairs of legs and no antennae (e.g. spider, scorpion, harvestman, mite).

Insects Arthropods with three pairs of legs, and in most cases wings. The body consists of three regions: the head, with eyes, mouth, and antennae; the thorax, to which legs and wings are attached; and the abdomen (e.g. grasshopper, housefly, butterfly, beetle).

Crustaceans

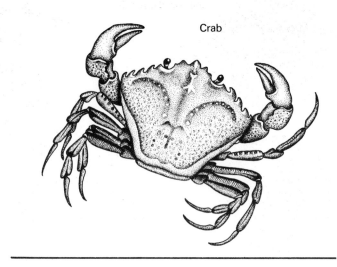

Crab

Arachnids

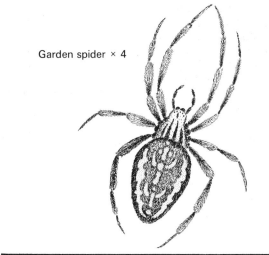

Garden spider × 4

Insects

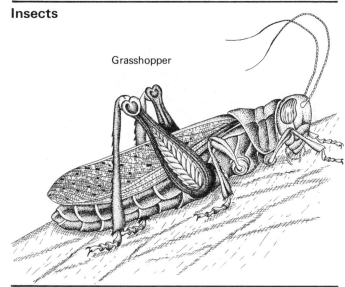

Grasshopper

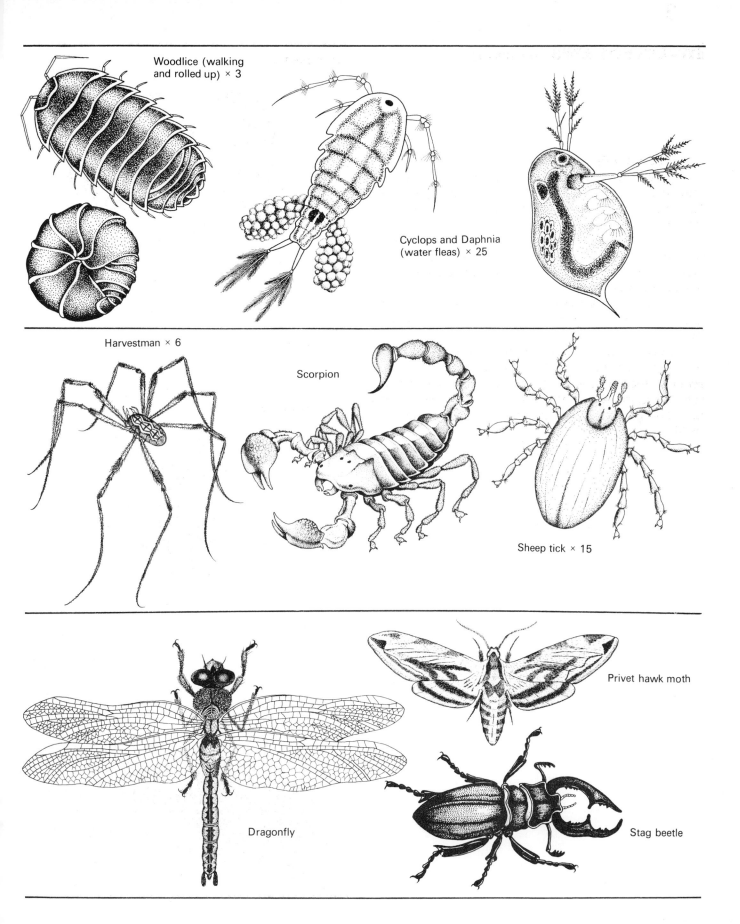

Woodlice (walking and rolled up) × 3

Cyclops and Daphnia (water fleas) × 25

Harvestman × 6

Scorpion

Sheep tick × 15

Privet hawk moth

Dragonfly

Stag beetle

4

Invertebrates *(continued)*

Arthropods *(continued)*

Myriapods Arthropods with a long body consisting of many segments. Millipedes have two pairs of legs on each segment; centipedes have one pair of legs on each segment.

Molluscs

Molluscs have a soft body enclosed in one or two shells. Some have a coiled shell and move about on a soft slimy 'foot' (e.g. snail). Some have two shells (e.g. mussel, oyster), and in some the shell is inside the body (e.g. octopus, squid, slug).

Echinoderms

Echinoderms live in the sea. They have a thick spiny skin, and the body is usually divided into five regions. A starfish, for instance, has five 'arms'. Echinoderms move by means of tiny tubular feet which end in a sucker (e.g. starfish, brittlestar, sea urchin, sea cucumber).

Myriapods

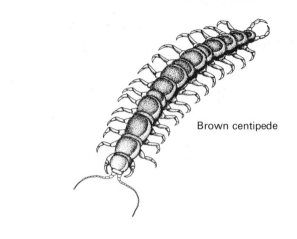

Brown centipede

Molluscs

Snail

Echinoderms

Starfish

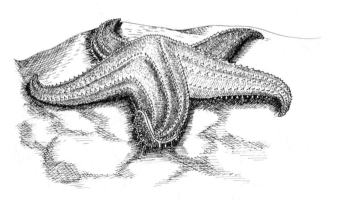

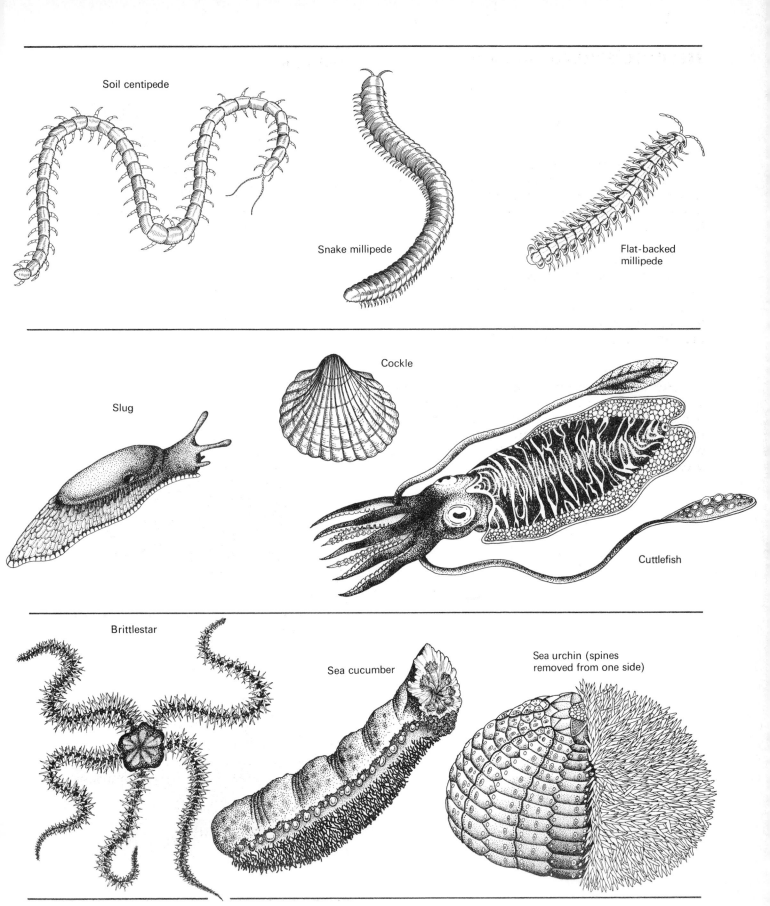

Soil centipede

Snake millipede

Flat-backed
millipede

Slug

Cockle

Cuttlefish

Brittlestar

Sea cucumber

Sea urchin (spines
removed from one side)

9

Vertebrates
(animals with backbones)

Fish

Fish live in water, have a streamlined body covered with scales, breathe through gills, and move by means of fins. Sharks, dogfish, and rays have a skeleton made of gristle (cartilage). The majority of fish have a skeleton of bone, and a swim bladder (e.g. cod, eel, sea-horse).

Amphibia

Amphibia have a moist skin without scales. They can live both on land and in water. They lay eggs covered with jelly in water. The eggs develop into larvae called tadpoles, which have gills and a tail (e.g. frog, toad, newt, salamander).

Reptiles

Reptiles have a dry scaly skin. Most snakes and lizards live on land, but terrapins, crocodiles, alligators, and some turtles live in water. Reptiles breathe with lungs, and lay eggs with tough leathery shells.

Fish

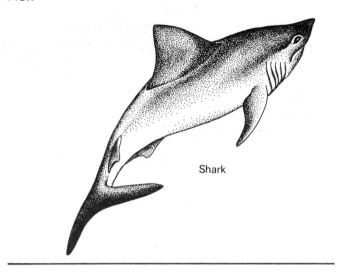

Shark

Amphibia

Toad

Reptiles

Giant tortoise

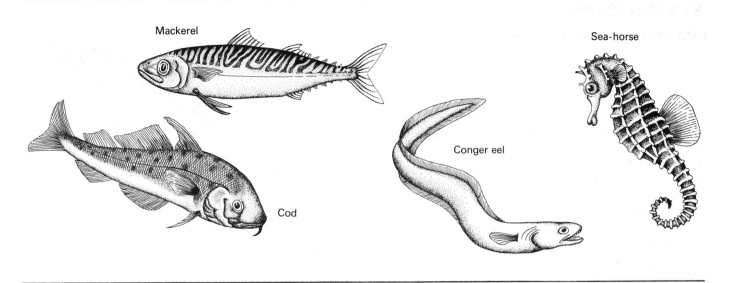

Mackerel

Sea-horse

Cod

Conger eel

Frog

Great crested newt

Marbled salamander

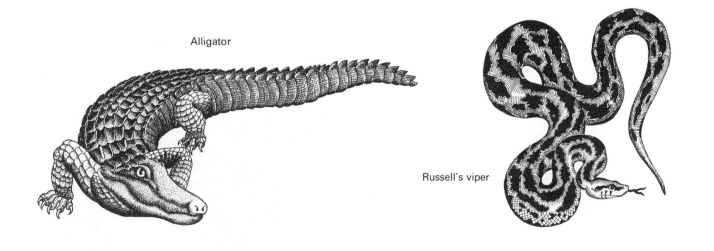

Alligator

Russell's viper

Vertebrates *(continued)*

Birds

Birds have a constant warm body temperature, feathers, and wings, which are front limbs adapted as organs of flight, (e.g. sparrow, seagull, duck). Some types cannot fly (e.g. penguin, kiwi).

Mammals

Mammals have a constant warm body temperature and a hairy skin. Female mammals suckle their young with milk from mammary glands, or breasts. Some of the main types of mammals are:

Monotremes Primitive egg-laying mammals (e.g. duck-billed platypus).

Marsupials Mammals which carry their young in a pouch on the abdomen (e.g. kangaroo, koala).

Primates The most advanced mammals. They have large brains, eyes positioned at the front of the head, and well developed hands capable of gripping, carrying, and manipulating objects (e.g. apes, monkeys, humans).

Birds

Eagle

Mammals

Duck-billed platypus

Kangaroo

Seagull

Sparrow

Duck

Penguin

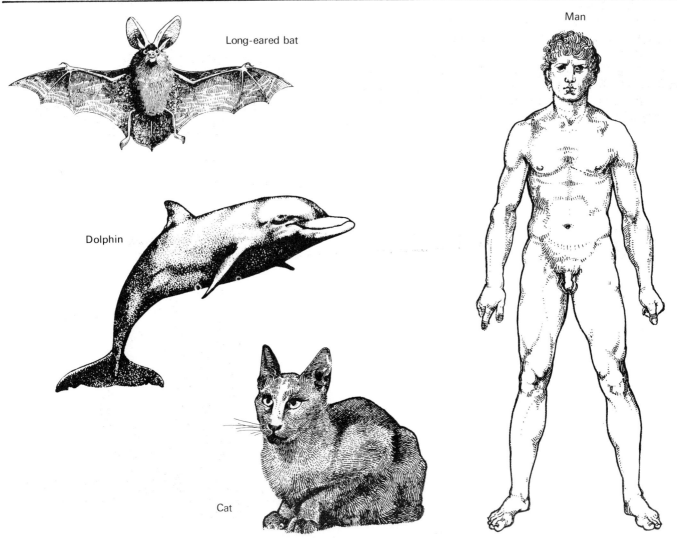

Long-eared bat

Man

Dolphin

Cat

Non-flowering plants

Mosses and liverworts

These plants do not have true roots, stems, or leaves, but they often have structures resembling these parts of higher plants. Many live in damp places, but even those from dry climates depend on water for reproduction, because they have swimming sperms. The fertilized 'egg', or ovum, develops into a capsule which bursts, releasing spores. These develop into new plants. Mosses have structures resembling roots, stems, and leaves (e.g. common cord moss, swan's neck thread moss). Liverworts are flat and leaf-like (e.g. wide-nerved liverwort).

Ferns and horsetails

These plants have a true root, stem, and leaves, but no flowers. They reproduce by means of spores produced from tiny capsules in clusters on the under-surfaces of the leaves. A spore develops into a leaf-like structure, resembling a liverwort, in which sexual reproduction occurs. The fertilized ovum develops into a new fern plant (e.g. male fern, hart's-tongue fern, polypody). Horsetail ferns produce spores from cones.

Conifers

These plants produce seeds inside a cone (e.g. pine, fir, spruce).

Mosses and liverworts

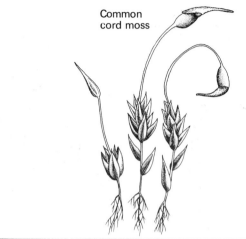

Common cord moss

Ferns

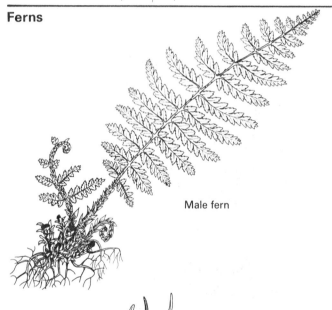

Male fern

Conifers

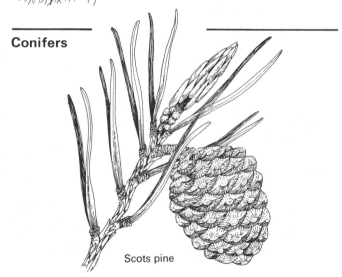

Scots pine

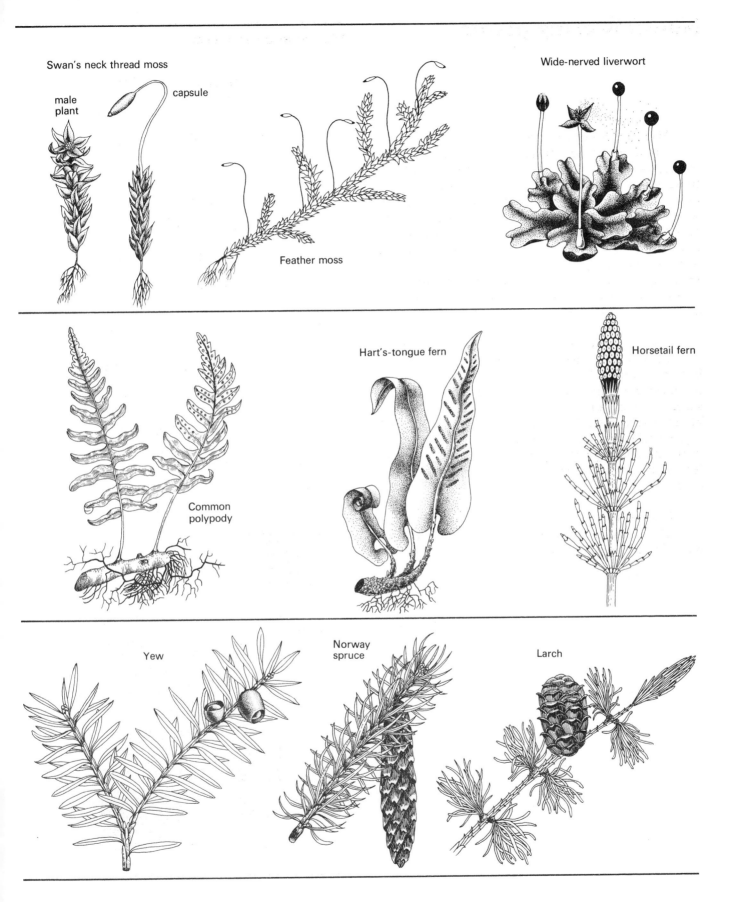

Swan's neck thread moss

male plant

capsule

Feather moss

Wide-nerved liverwort

Common polypody

Hart's-tongue fern

Horsetail fern

Yew

Norway spruce

Larch

Flowering plants

These plants produce flowers containing reproductive organs. Male reproductive organs, called stamens, produce pollen grains containing male sex cells. These fertilize ova (female sex cells) inside an ovary. The ovaries form part of the female reproductive organs or carpels. Each fertilized ovum develops into a seed, and is contained in a fruit which develops from the ovary wall.

Monocotyledons
These flowering plants germinate with only one seed leaf (cotyledon), and their leaves have parallel veins (e.g. grass, tulip, lily, banana).

Dicotyledons
This is the largest group of flowering plants. The seeds germinate with two seed leaves, and the leaves have a network of veins (e.g. oak, buttercup, dead-nettle, sweet pea).

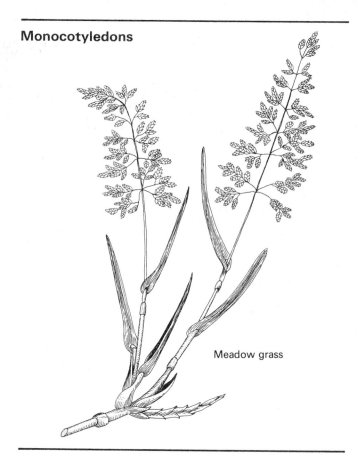

Meadow grass

Dicotyledons

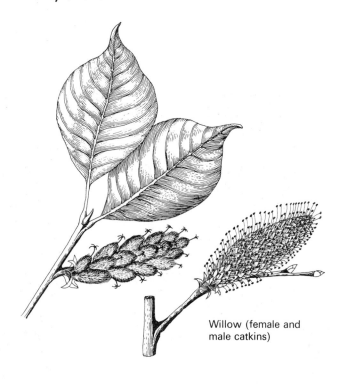

Willow (female and male catkins)

16

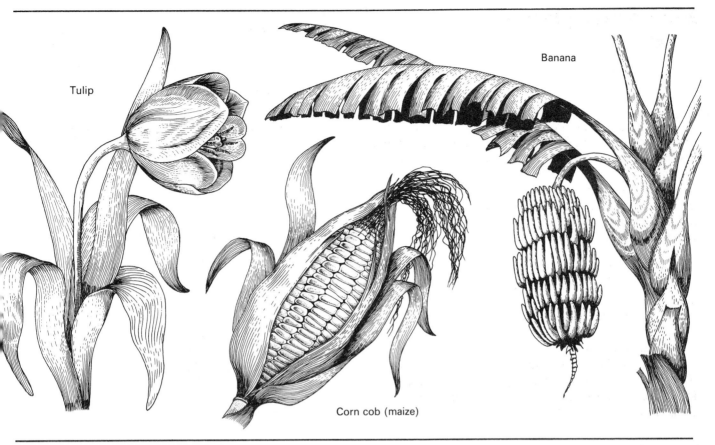

Tulip

Banana

Corn cob (maize)

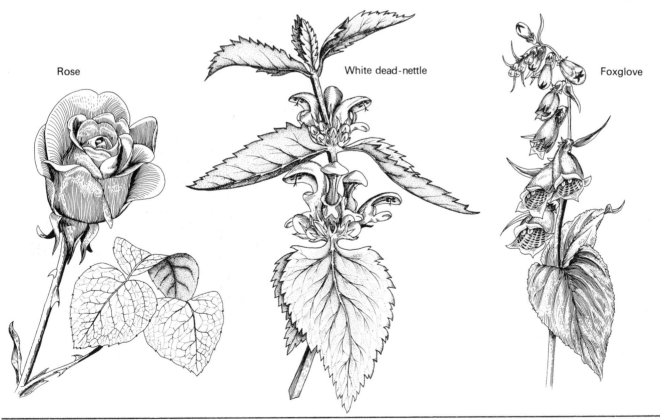

Rose

White dead-nettle

Foxglove

17

9

Keys and how to use them

In biology a key is a device used to identify living things. There are many types of key, but the simplest consist of a list of descriptions arranged in numbered pairs (Fig. 9.2). To identify an organism look at the first pair of descriptions and choose the one which fits. The key will then either say what the organism is, or it will give the number of another pair of descriptions. Go on to the next pair of descriptions if necessary, and repeat this process until the specimen can be identified.

For example, Animal A in Figure 9.2 is identified in the following way. Study the drawing and read the first pair of descriptions. It is clear that the animal does not live in a tube. Therefore go on to the second set of descriptions. Animal A does not have two trumpet-shaped breathing tubes on its head, so go to the third set of descriptions. The animal's tail has prongs, so look at the fourth pair of descriptions. Animal A has a two-pronged tail, which means it must be a stonefly nymph.

Identify the other animals in Figure 9.2, starting each time with the first set of descriptions and working through the list until you find out what the animal is called. Copy Figure 9.1 on to a separate piece of paper. Fill in the remaining columns, recording each stage of the identification process and the animal's name, in the same way as Animal A.

Animal	Stages	Name
A	1, 2, 3, 4	Stonefly nymph
B		
C		
D		
E		
F		
G		
H		

Fig. 9.1 Copy and complete this chart by identifying the animals in Figure 9.2. Record both the animal's name and the number of each stage necessary to identify it. (Do not write in the textbook.)

Fig. 9.2 (right) Key to insect nymphs, larvae, and pupae

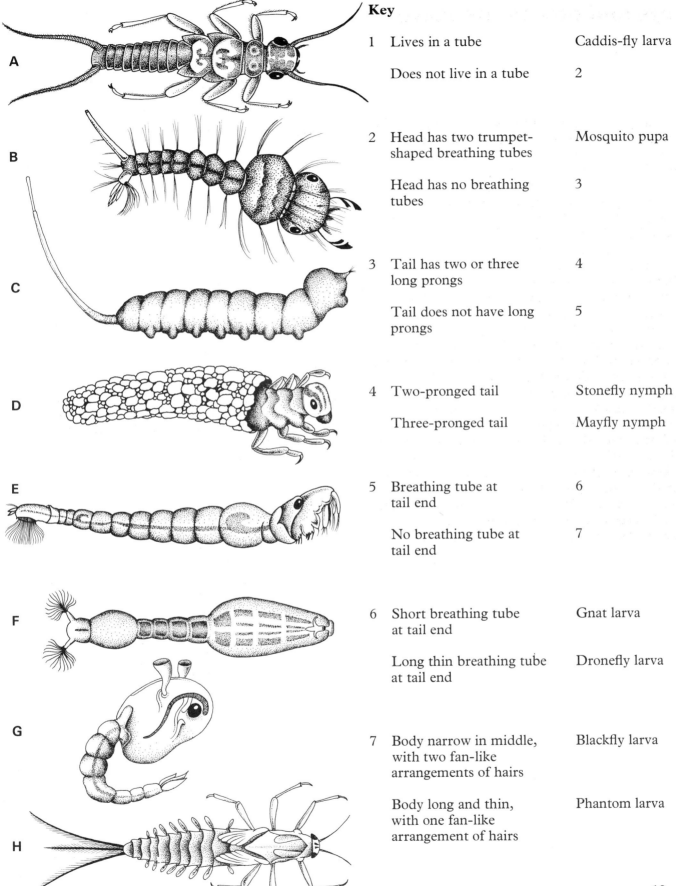

Key

1	Lives in a tube	Caddis-fly larva
	Does not live in a tube	2
2	Head has two trumpet-shaped breathing tubes	Mosquito pupa
	Head has no breathing tubes	3
3	Tail has two or three long prongs	4
	Tail does not have long prongs	5
4	Two-pronged tail	Stonefly nymph
	Three-pronged tail	Mayfly nymph
5	Breathing tube at tail end	6
	No breathing tube at tail end	7
6	Short breathing tube at tail end	Gnat larva
	Long thin breathing tube at tail end	Dronefly larva
7	Body narrow in middle, with two fan-like arrangements of hairs	Blackfly larva
	Body long and thin, with one fan-like arrangement of hairs	Phantom larva

19

Structure of cells

Cells can be described as the 'building-blocks' of life. Like the bricks which make up a wall, cells are the basic parts out of which all but the simplest living things are made. But bricks are non-living, identical in shape, and quite large; whereas cells are alive, vary enormously in shape, and are microscopic in size.

Cell structure

All cells have the following parts in common:

Cell membrane The cell membrane is an extremely thin 'skin' which forms the outer boundary of a cell. This is where a cell absorbs useful substances and removes its waste materials. A cell membrane allows only certain substances to enter and leave a cell. It is therefore said to be a **semi-permeable** membrane.

Cytoplasm Cytoplasm is all the living material in a cell except the nucleus. It is a jelly-like material consisting of hundreds of chemicals involved in hundreds of reactions. Some reactions occur in specialized regions of cytoplasm such as **mito-chondria** (Fig. 10.2) in which respiration takes place, or the **endoplasmic reticulum**, which makes proteins.

The nucleus At least one nucleus is found in the cells of all animals, plants, and protists. The nucleus controls the whole cell, mainly by regulating the manufacture of proteins. It also plays an essential part in cell division.

Differences between plant and animal cells

Plant cells (Figs. 10.1 and 10.3) All plant cells are enclosed in a tough layer of **cellulose**. The cytoplasm of a mature plant cell contains one or more large, permanent spaces called **vacuoles**. These are filled with a liquid called cell sap, and are lined with a semi-permeable membrane. The green parts of a plant owe their colour to the presence of **chlorophyll**, which is contained within tiny objects in the cytoplasm called **chloroplasts**. Chlorophyll absorbs light energy, which is necessary for photosynthesis.

Animal cells Animal cells have no cellulose cell wall, and never contain chlorophyll. Small, temporary vacuoles often occur in their cytoplasm. There is a greater variety of shape and function among animal cells than among plant cells.

A An animal cell

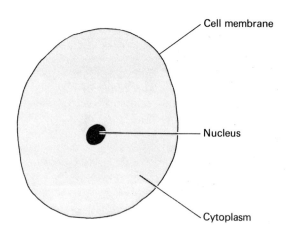

B A plant cell

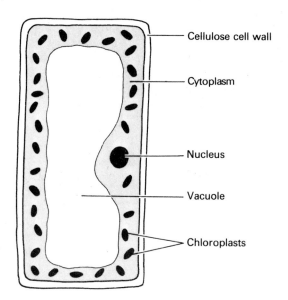

Fig. 10.1 Diagrams showing basic structure of animal and plant cells

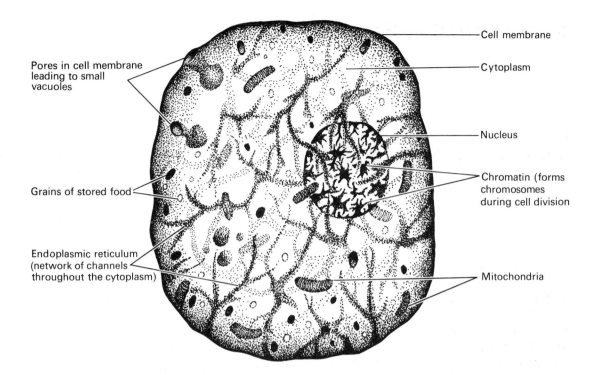

Cell membrane

Cytoplasm

Pores in cell membrane
leading to small
vacuoles

Nucleus

Chromatin (forms
chromosomes
during cell division

Grains of stored food

Endoplasmic reticulum
(network of channels
throughout the cytoplasm)

Mitochondria

Fig. 10.2   An unspecialized animal cell (magnified about one
million times)

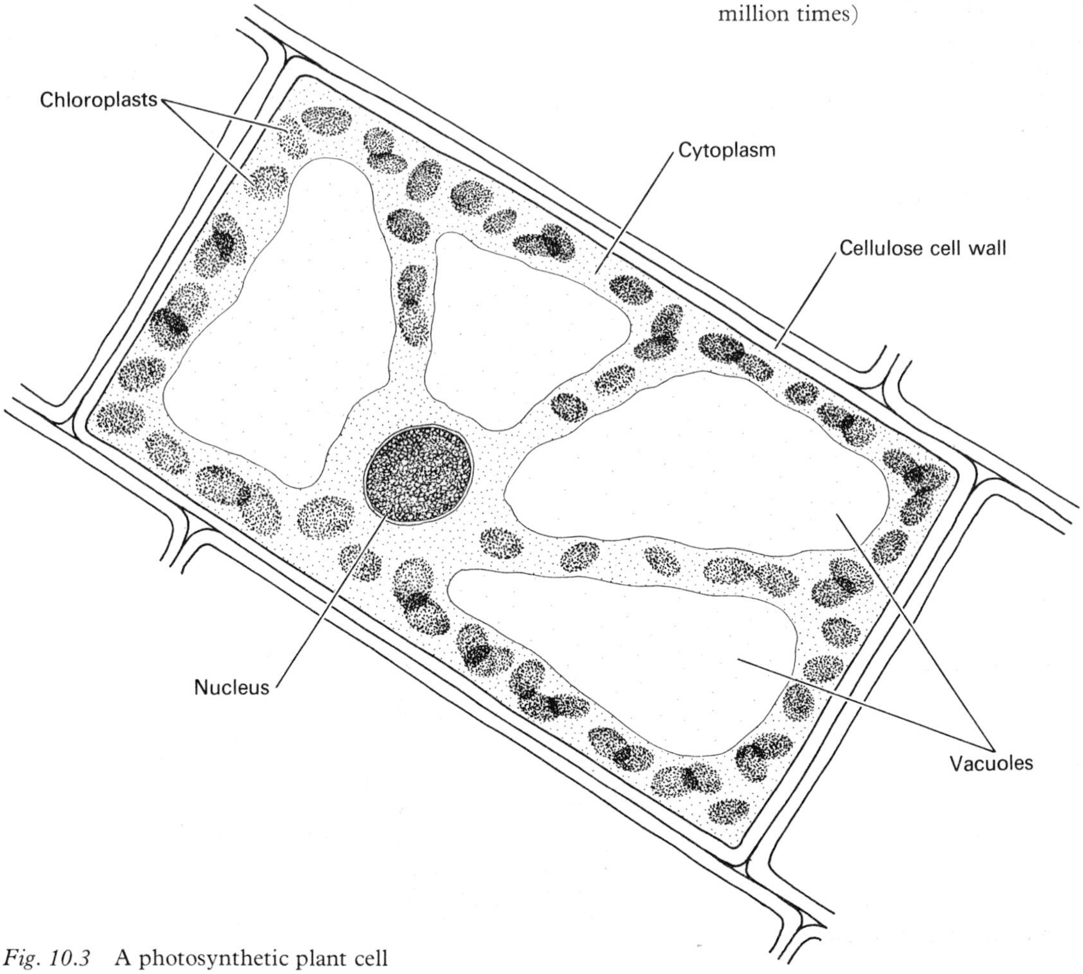

Chloroplasts

Cytoplasm

Cellulose cell wall

Nucleus

Vacuoles

Fig. 10.3   A photosynthetic plant cell

11

Cells, tissues, and organs

Multicellular organisms (those made of many cells) usually consist of many different parts. These parts consist of cells with special features that enable them to perform a particular task efficiently. In other words, the cells are **specialized**. Groups of specialized cells are called **tissues**. Muscle tissue, for example, is made of specialized cells which can contract and move the body. Nervous tissue consists of cells which can carry nerve impulses. Photosynthetic tissue in plants consists of cells filled with chloroplasts. These absorb sunlight energy which is used in other parts of the cell to drive the chemical reactions of photosynthesis.

An **organ** is made up of several different tissues, each of which contributes to the functions of the organ as a whole. The heart and liver are animal organs, and leaves and roots are plant organs.

Several organs working in conjunction form an **organ system**. The circulatory system, composed of the heart, blood, and blood vessels, is an example.

Division of labour

The way in which various parts of multicellular organisms are specialized for one particular function is an example of division of labour. This means that the *labour* involved in maintaining the organism's life processes is *divided* among specialized parts. The advantage of division of labour is that it is highly efficient. Just as a man trained to do one particular job is likely to be better at it than a jack-of-all-trades, so a specialized body tissue performs its function more efficiently than a group of unspecialized cells with many different functions.

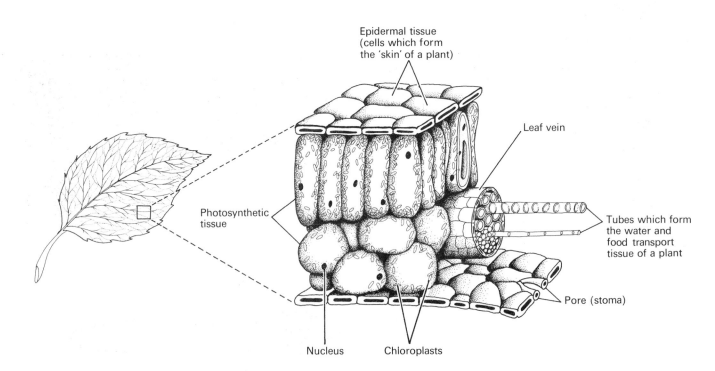

Fig. 11.1 Plant cells and tissues

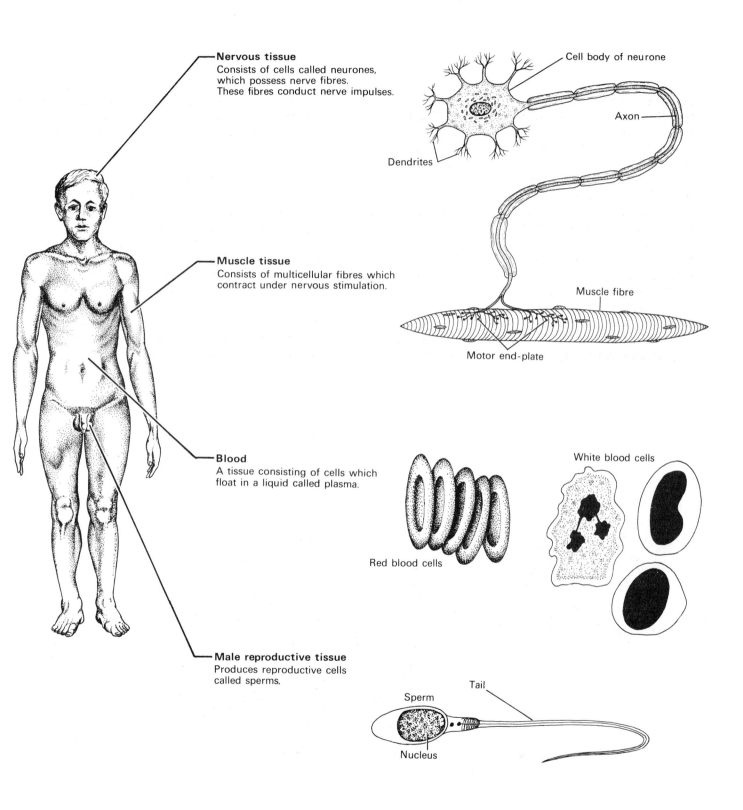

Nervous tissue
Consists of cells called neurones, which possess nerve fibres. These fibres conduct nerve impulses.

Cell body of neurone

Axon

Dendrites

Muscle tissue
Consists of multicellular fibres which contract under nervous stimulation.

Muscle fibre

Motor end-plate

Blood
A tissue consisting of cells which float in a liquid called plasma.

White blood cells

Red blood cells

Male reproductive tissue
Produces reproductive cells called sperms.

Tail

Sperm

Nucleus

Fig. 11.2 Human cells and tissues

23

12

Cell division

Plants and animals are usually composed of millions of cells when they are fully grown, yet they all begin life from only one cell: a fertilized egg cell or **zygote**. The zygote divides over and over again, producing 'daughter' cells which eventually become specialized into all the body tissues.

Zygotes of most animals are very similar. It would be hard to tell the difference between the zygote of a cat or a human, for example. Why then do cat zygotes produce only cats and not humans? What makes a zygote turn into the same species of organism as its parents?

The answer is that the nucleus of a zygote contains all the 'instructions', in chemical form, to build one particular type of organism. These instructions are contained within **chromosomes**, which make up the bulk of a nucleus.

Chromosomes are only visible when a cell divides. At first they appear as long fine threads, but these become shorter by coiling up (Fig. 12.1).

The type of cell division which eventually produces an adult organism from a zygote is called **mitosis**.

Mitosis is responsible for most forms of growth. During mitosis a cell divides in such a way that its daughter cells have the same number and type of chromosomes as the cell which produced them. Every cell passes on an exact copy of its 'building instructions' to each of its daughters, and as the organism grows these 'instructions' determine the characteristic tissues and organs of its species.

During the first stage of mitosis the chromosomes become short and thick, and each consists of two threads (Fig. 12.2A). Next, the chromosomes move to the equator (middle) of the cell where they become attached to fine fibres called the **spindle** (Fig. 12.2B). Then each chromosome separates into two parts, which move away from each other towards opposite ends of the cell (Fig. 12.2C). The spindle fibres probably contract, pulling the chromosomes apart. Next, the two groups of chromosomes gather at opposite ends of the cell. Each group forms a nucleus, and the cell begins to divide into two (Fig. 12.2D). Finally, the cell divides completely into two daughter cells (Fig. 12.2E).

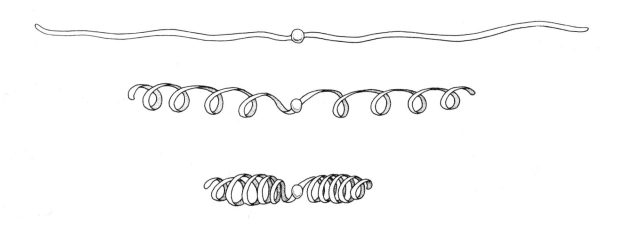

Fig. 12.1 Diagram showing how chromosomes become shorter and thicker by coiling up

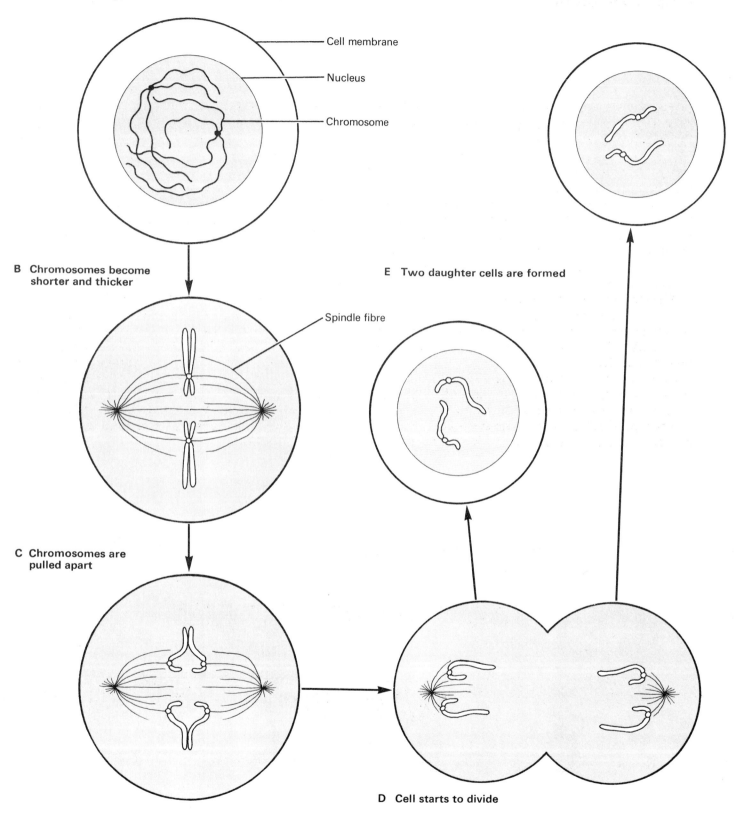

A Chromosomes appear

Cell membrane

Nucleus

Chromosome

B Chromosomes become shorter and thicker

Spindle fibre

E Two daughter cells are formed

C Chromosomes are pulled apart

D Cell starts to divide

Fig. 12.2   Diagram of cell division by mitosis

13

Types of support

Land animals are supported against the force of gravity by a skeleton. There are two main types of skeleton: exoskeletons and endoskeletons.

Exoskeletons

Arthropods, such as crustaceans (e.g. crayfish and lobsters), and insects (e.g. bees and flies), are supported by a layer of tough, rigid material called the **cuticle**, which forms on the outsides of their bodies. A cuticle is an example of an exoskeleton. Arthropod exoskeletons are made up of thick, rigid plates and tubes held together at the joints by a thinner, more flexible type of cuticle (Fig. 13.2), which enables the animal to move. The outer surface of the cuticle is coated with a thin layer of wax which makes it waterproof. Below the wax is a tough but light material called **chitin**. In crustaceans chitin is mixed with chalk. This makes their exoskeletons much heavier and more rigid than those of insects which do not contain chalk.

An arthropod exoskeleton not only supports the body; it also protects the soft inner parts from physical damage, stops them drying up, and prevents dirt and germs from entering.

Endoskeletons

Vertebrates (fish, amphibia, reptiles, birds, and mammals) are supported by an internal skeleton of bones, known as an endoskeleton. The main feature of this type of skeleton is a long flexible rod of small bones called **vertebrae**, which forms the backbone or **vertebral column**.

Bones consist of living cells embedded in hard mineral substances (mainly calcium compounds). These minerals give bones their strength and shape. Bones are held together at the joints by tough, flexible fibres called **ligaments**.

Bones not only support the body. They also provide protection. The skull, for instance, protects the brain, and in mammals the rib cage protects the lungs, heart, and main blood vessels. Bones also form a system of rods and levers which can be moved in various ways by the muscles.

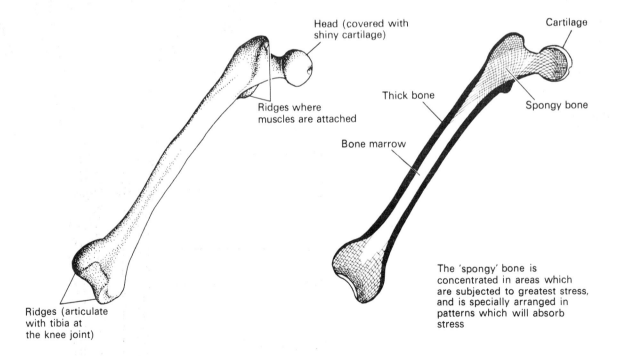

Head (covered with shiny cartilage)

Ridges where muscles are attached

Ridges (articulate with tibia at the knee joint)

Cartilage

Thick bone

Spongy bone

Bone marrow

The 'spongy' bone is concentrated in areas which are subjected to greatest stress, and is specially arranged in patterns which will absorb stress

Fig. 13.1   Structure of a bone (human femur)

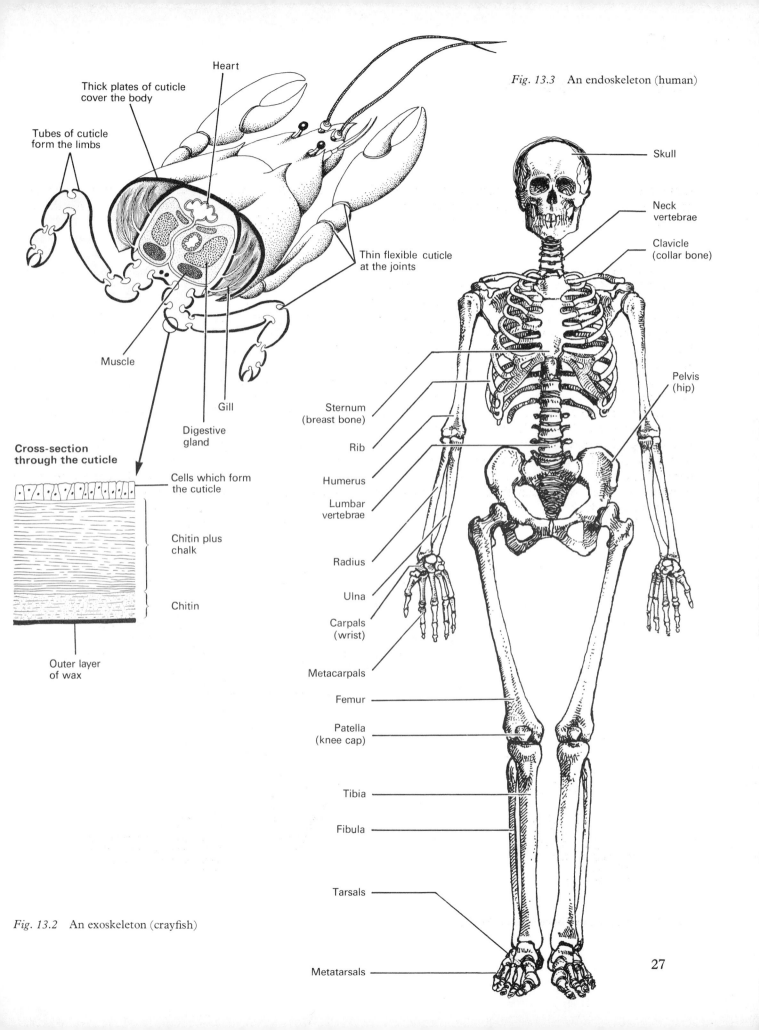

Heart

Thick plates of cuticle
cover the body

Tubes of cuticle
form the limbs

Fig. 13.3 An endoskeleton (human)

Thin flexible cuticle
at the joints

Muscle

Gill

Digestive
gland

Cross-section
through the cuticle

Cells which form
the cuticle

Chitin plus
chalk

Chitin

Outer layer
of wax

Fig. 13.2 An exoskeleton (crayfish)

Skull

Neck
vertebrae

Clavicle
(collar bone)

Pelvis
(hip)

Sternum
(breast bone)

Rib

Humerus

Lumbar
vertebrae

Radius

Ulna

Carpals
(wrist)

Metacarpals

Femur

Patella
(knee cap)

Tibia

Fibula

Tarsals

Metatarsals

27

Joints in the human skeleton

Joints occur wherever two or more bones touch. In some joints the bones are fastened firmly together by fibres so that no movement is possible. These are called **fixed joints** (Fig. 14.1A). Fixed joints occur between the flattened bones which form the roof of the skull. In other joints the bones move slightly against a pad of gristle (cartilage) situated between them (Fig. 14.1B). These are **slightly moveable joints** and occur between the vertebrae of the backbone (Fig. 14.2B). The cartilage also absorbs shock which would otherwise pass up the backbone to the skull during running and jumping. Most joints are **freely moveable** or **synovial joints** (Fig. 14.1C). These allow free movement in one or more directions. Where the bones of these joints rub together they are covered with a layer of shiny, slippery cartilage. The joint is lubricated by a liquid called **synovial fluid**, sealed in by the **synovial membrane** which surrounds the whole joint. The bones are held together, but allowed to move freely, by fibres called **liga-ments**. The type of movement possible at a synovial joint depends on the shape of the bones at the point where they rub together.

In **hinge joints**, such as the elbows (Fig. 14.2A) and knees, movement can occur in only one direction, like the hinge of a door.

In **ball-and-socket joints**, such as the hip (Fig. 14.2C) and shoulder, the rounded head of one bone fits into a cup-shaped socket in another. These joints allow movement in all directions.

In **pivot joints** one bone twists against another. There is a pivot joint in the elbow (Fig. 14.2A) where the radius bone twists against the ulna. This enables the hands to perform 'screwdriver' movements, that is, the palms can either face upwards or downwards.

In **sliding** or **gliding joints** the surfaces which rub together are flat. Examples occur between the vertebrae where two flat-surfaced projections from each vertebra move against each other when the backbone bends (Fig. 14.2B).

A Fixed or immoveable joint   **B Slightly moveable joint**   **C Freely moveable (synovial) joint**

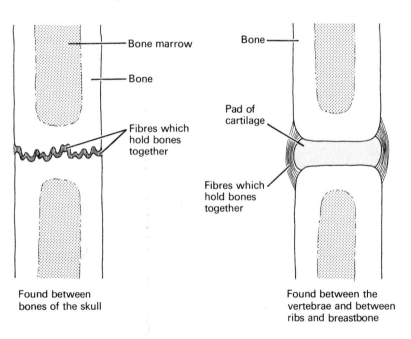
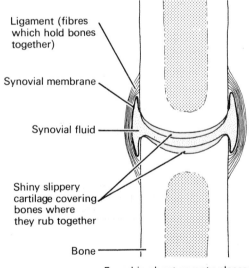

Fig. 14.1 Diagrams of the main types of joint

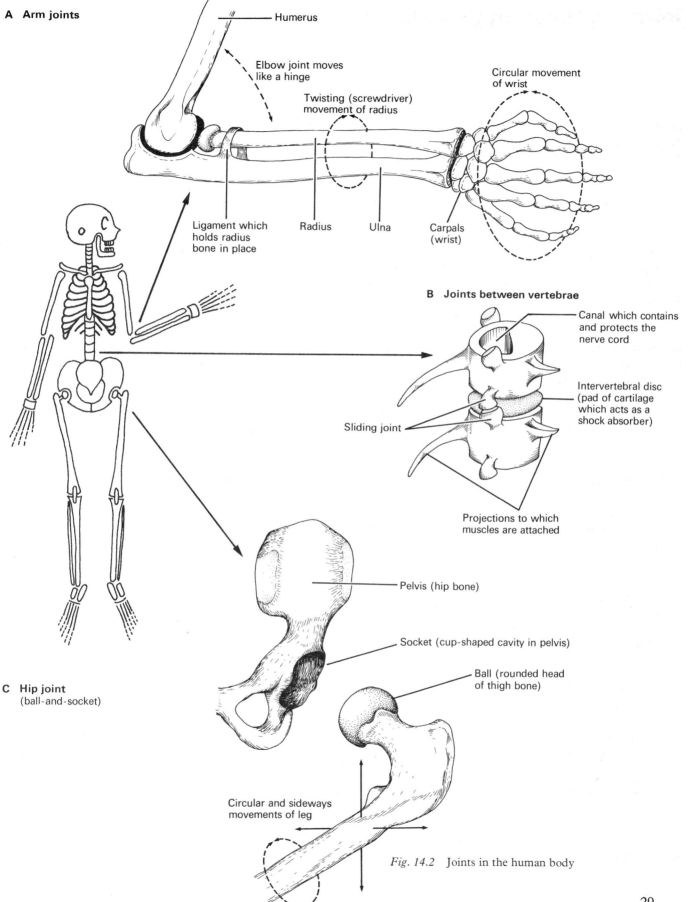

A Arm joints

Humerus

Elbow joint moves like a hinge

Twisting (screwdriver) movement of radius

Circular movement of wrist

Ligament which holds radius bone in place

Radius

Ulna

Carpals (wrist)

B Joints between vertebrae

Canal which contains and protects the nerve cord

Intervertebral disc (pad of cartilage which acts as a shock absorber)

Sliding joint

Projections to which muscles are attached

Pelvis (hip bone)

Socket (cup-shaped cavity in pelvis)

Ball (rounded head of thigh bone)

C Hip joint (ball-and-socket)

Circular and sideways movements of leg

Fig. 14.2 Joints in the human body

29

15

Muscles and movement

Muscles are the 'meat' of the body, and in mammals make up 40–50 per cent of body weight. Muscles which are attached to the skeleton are called **skeletal** or **voluntary** muscles. These can be moved at will. In mammals, muscles are attached to the skeleton by fibres called **tendons** (Fig. 15.3), and in arthropods (e.g. insects) by internal projections from the cuticle (Fig. 15.2).

Muscles cause movement by contracting, that is, they become shorter. As this happens they pull against the skeleton, making it bend at a joint. The end of a muscle closest to the joint is called the **insertion** of the muscle. The insertion moves during muscular contraction while the opposite end, called the **origin**, remains fixed as the anchorage point of the muscle.

There are two sets of muscles at each joint. One set, called the **extensor muscles**, are attached in such a way that when they contract the limb is straightened (Fig. 15.1A). On the opposite side of the joint there are **flexor muscles**, which bend the joint (Fig. 15.1B). Since extensor and flexor muscles pull a joint in opposite directions they are called an **antagonistic system**.

During movements bones act as **levers**. The point where one lever moves against another is called the **fulcrum**. In a skeleton a joint acts as a fulcrum (Fig. 15.1). The force which moves a lever is known as **effort**. In a skeleton effort is supplied by muscles. All levers carry or move a **load** of some kind. The load which the skeleton supports is the weight of the body, and any additional weight being carried.

A Action of extensor muscle

Effort
(contraction of
extensor muscle)

Load (arm
and hand)

Fulcrum
(elbow joint)

B Action of flexor muscle

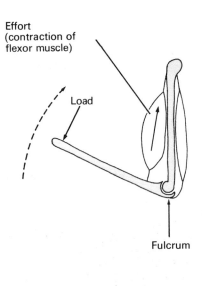

Effort
(contraction of
flexor muscle)

Load

Fulcrum

Fig. 15.1 Diagram showing how antagonistic muscles move the elbow joint

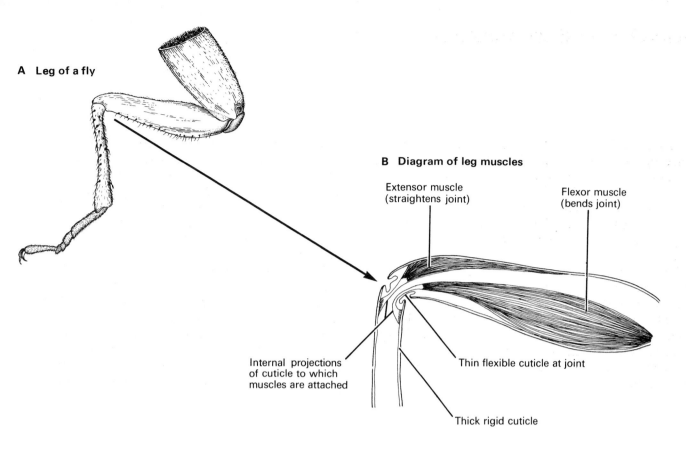

A Leg of a fly

B Diagram of leg muscles

Extensor muscle
(straightens joint)

Flexor muscle
(bends joint)

Internal projections
of cuticle to which
muscles are attached

Thin flexible cuticle at joint

Thick rigid cuticle

Fig. 15.2 Antagonistic muscles in an insect's leg

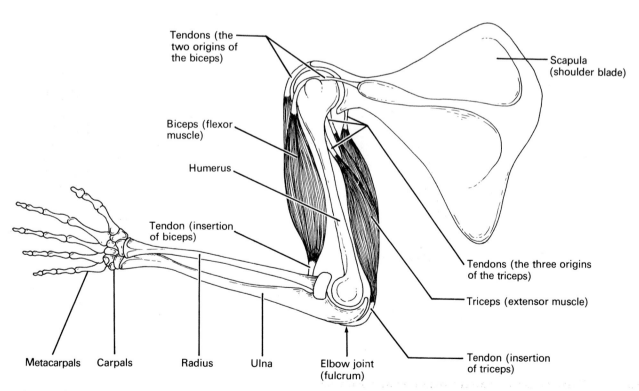

Tendons (the
two origins of
the biceps)

Scapula
(shoulder blade)

Biceps (flexor
muscle)

Humerus

Tendon (insertion
of biceps)

Tendons (the three origins
of the triceps)

Triceps (extensor muscle)

Tendon (insertion
of triceps)

Metacarpals   Carpals   Radius   Ulna   Elbow joint
(fulcrum)

Fig. 15.3 Antagonistic muscles in the human arm

16

Bird flight

A bird has two main features which help it to overcome the force of gravity and fly through the air. First, its wings, when seen in cross-section, have a curved shape called an **aerofoil** (Fig. 16.1). This shape produces a force called **lift** as a wing moves through the air. Second, when a bird flaps its wings up and down further lift is produced which drives the bird forwards through the air.

How an aerofoil produces lift

As an aerofoil moves through the air its shape forces the air to move faster across its upper surface than across its under surface. This is important because fast moving air always has a *lower* pressure than slow moving air. Consequently, as a bird moves through the air it is lifted upwards by low pressure air above its wings, and at the same time it is pushed upwards by high pressure air underneath its wings.

The main aerofoil region of a bird's wings is the part to which the **secondary feathers** are attached (Fig. 16.2). The **primary feathers** at the wing tips provide both lift and forward thrust during flapping movements.

How wing movements produce lift and forward thrust

Figure 16.3 shows that wings do not simply move up and down. During the **upstroke** the front edge of a wing is higher than its trailing edge. In this position the wing is lifted upwards and backwards by the build-up of air pressure underneath it. This movement is helped by the feathers separating like the slats of a venetian blind and allowing air to pass through the wings.

During the **downstroke** the feathers close up and muscles pull the wing downwards and forwards. The front edge of the wing is lower than the trailing edge, and so the wing pushes the air backwards. This drives the bird forwards.

Birds have other features which help them to fly efficiently: they have a streamlined shape, and their bones are hollow and filled with air. This latter feature makes the skeleton extremely light, but its strength is not reduced.

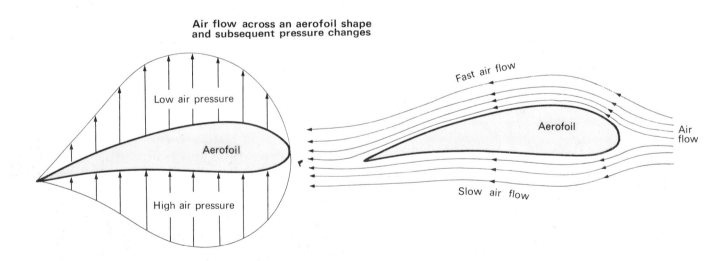

Fig. 16.1 Diagram showing how an aerofoil generates lift

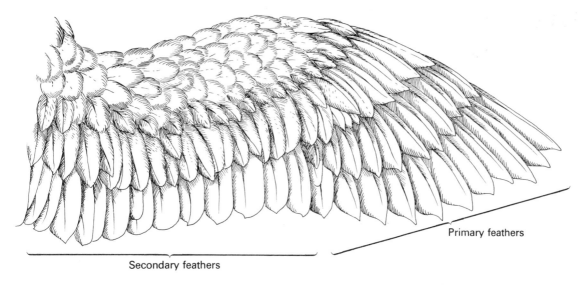

Fig. 16.2 Arrangement of feathers on a bird's wing

Secondary feathers

Primary feathers

Downstroke
Front of wing is lower than back and air pressure closes feathers, making the wing airtight

Upstroke
Front of wing is higher than back and feathers separate, letting air pass through wing

Fig. 16.3 Positions of the wing during one wing-beat

17

Movement in fish

A fish swims by sweeping its tail fin from side to side. During these movements the tail fin pushes backwards against the water, and this pushes the fish forwards (Figs. 17.2 and 17.5).

The tail fin is moved by blocks of muscle situated on either side of the backbone and around the sides of the body (Fig. 17.1). The backbone is a long flexible rod made up of small bones called **vertebrae**. The muscles contract alternately on each side of the body, bending the backbone first one way and then the other. These muscular contractions and bending movements sweep the tail fin from side to side.

The dorsal, ventral, pelvic, and pectoral fins (Fig. 17.3) control the direction of swimming movements.

The dorsal and ventral fins prevent the fish from **rolling** or 'corkscrewing' through the water, and from **yawing** or sideways movements. The pelvic and pectoral fins prevent **pitching**, or see-saw movements (Fig. 17.4).

Most fish have an air-filled space inside their bodies called a **swim bladder**. The amount of air in the swim bladder is increased or decreased until the fish is completely weightless and floats effortlessly in the water. The advantage of the swim bladder is that since the fish is weightless, it need use no energy in overcoming the force of gravity. Sharks and some other fish have no swim bladder, and they sink to the bottom if they stop swimming.

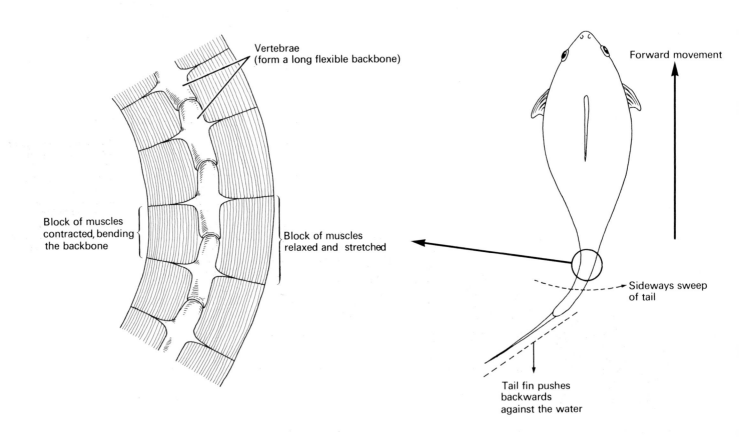

Fig. 17.1   Swimming muscles of a fish

Fig. 17.2   Diagram showing how the tail fin causes the fish to move forwards

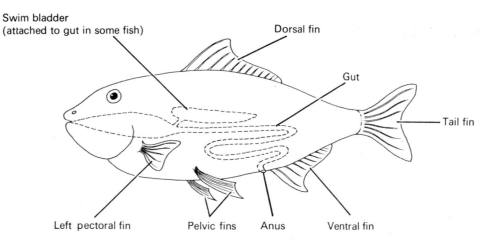

Fig. 17.3 Names of fins and position of swim bladder

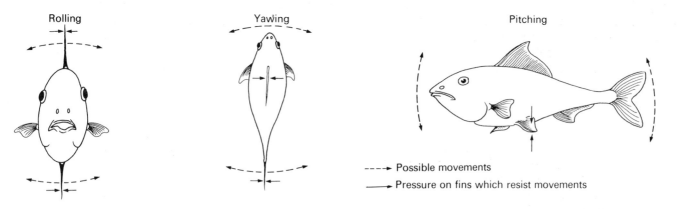

Fig. 17.4 Diagram showing how fins control body movements   Fig. 17.5 (below) Diagram showing how a fish swims

These are drawings of a fish viewed from above as it swims Note that as the tail sweeps sideways it presses against the water up to the moment when it changes direction and repeats the process

18

Photosynthesis

Unlike animals, plants do not get food by eating other living things: they make it themselves by means of a process called photosynthesis. During photosynthesis plants make a sugar called **glucose**. Glucose, plus many other chemicals absorbed from the soil, are then used as raw materials in the manufacture of all the substances which make up a plant.

Photosynthesis requires energy, carbon dioxide gas, and water. The energy comes from sunlight, and is absorbed by a green substance called **chlorophyll**. This is found mainly in the leaves inside microscopic objects called **chloroplasts**. Carbon dioxide comes from the air, and is absorbed through tiny pores called **stomata**. In most plants these are found mainly on the under-surfaces of the leaves. Water is absorbed from the soil by the roots. Water passes up the roots, stem, and into the leaves through narrow tubes called **xylem vessels**. In the leaves xylem vessels are contained in the mid-rib and the branching network of veins.

During photosynthesis sunlight energy is used to combine carbon dioxide with water. This produces glucose sugar, and oxygen gas is released into the air as a waste product:

$$\underset{\substack{\text{Carbon} \\ \text{dioxide}}}{6CO_2} + \underset{\text{Water}}{6H_2O} + \underset{\text{Energy}}{\text{Light}} \overset{\text{Chlorophyll}}{\longrightarrow} \underset{\text{Glucose}}{C_6H_{12}O_6} + \underset{\text{Oxygen}}{6O_2}$$

Photosynthesis in leaves takes place inside a layer of cells called the **mesophyll**. The uppermost region of the mesophyll consists of long cylindrical cells called the **palisade layer**. Cells of the palisade layer contain more chlorophyll than any other leaf cells. Below the palisade is the **spongy layer** of the mesophyll, so called because of the large air spaces between its cells. The spongy layer also contains chloroplasts and carries out photosynthesis.

The food (glucose) made during photosynthesis passes out of the leaves and is transported to all parts of the plant through tubes which make up a tissue called **phloem**.

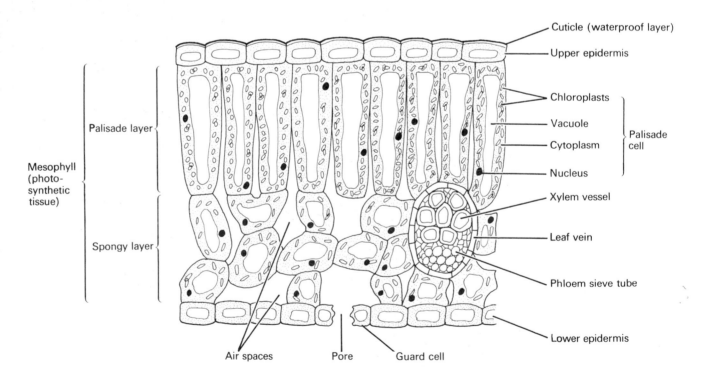

Fig. 18.1   Diagram of the structure of a leaf

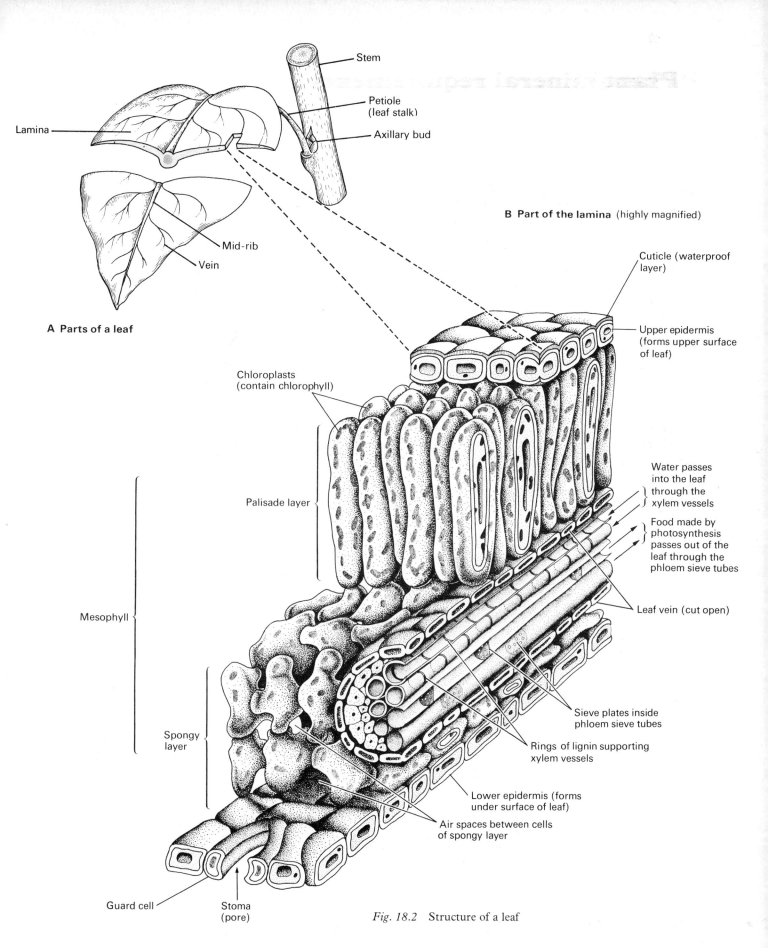

Lamina

Stem

Petiole
(leaf stalk)

Axillary bud

Mid-rib

Vein

A Parts of a leaf

B Part of the lamina (highly magnified)

Cuticle (waterproof
layer)

Upper epidermis
(forms upper surface
of leaf)

Chloroplasts
(contain chlorophyll)

Palisade layer

Water passes
into the leaf
through the
xylem vessels

Food made by
photosynthesis
passes out of the
leaf through the
phloem sieve tubes

Leaf vein (cut open)

Mesophyll

Sieve plates inside
phloem sieve tubes

Spongy
layer

Rings of lignin supporting
xylem vessels

Lower epidermis (forms
under surface of leaf)

Air spaces between cells
of spongy layer

Guard cell

Stoma
(pore)

Fig. 18.2   Structure of a leaf

Plant mineral requirements

Plants make sugar during photosynthesis, but they cannot live on sugar alone. In order to grow and make flowers and seeds they require proteins. Plants make their own proteins from sugar and minerals extracted from soil by their roots.

Certain minerals are called **major elements** because they are required by plants in fairly large quantities. The major elements are: nitrogen, phosphorus, sulphur, potassium, calcium, and magnesium. Other minerals are called **trace elements**. These are also necessary, but only in minute quantities. Some of the trace elements are: manganese, copper, zinc, iron, boron, and molybdenum.

Figure 19.1 illustrates an experiment to study plant mineral requirements. A number of *Tradescantia* cuttings of about the same length are placed with their stems in solutions which lack just *one* of the minerals essential for healthy growth. Over a period of about a month the growth of these cuttings is compared with other cuttings grown in a solution containing *all* the minerals that plants require. *Tradescantia* cuttings are used because they root quickly. Since cuttings possess no stored food they depend entirely upon photosynthesis and minerals in solution right from the start of the experiment.

Plants grown in a solution lacking one mineral soon show symptoms of ill health, when compared with cuttings grown in solutions containing all the minerals a plant needs. These are called **mineral deficiency symptoms**, and vary according to which mineral is missing from the solution (Fig. 19.2).

In agriculture, mineral deficiency symptoms tell farmers which minerals they must add to their soil to obtain healthy plants.

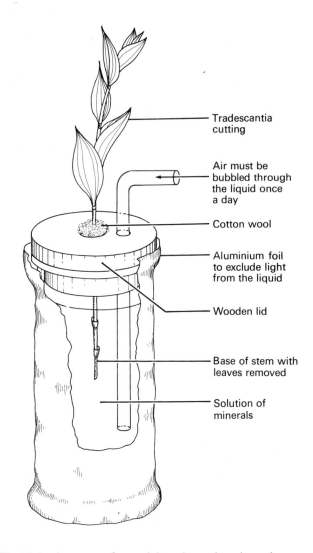

Tradescantia cutting

Air must be bubbled through the liquid once a day

Cotton wool

Aluminium foil to exclude light from the liquid

Wooden lid

Base of stem with leaves removed

Solution of minerals

Fig. 19.1 Apparatus for studying plant mineral requirements

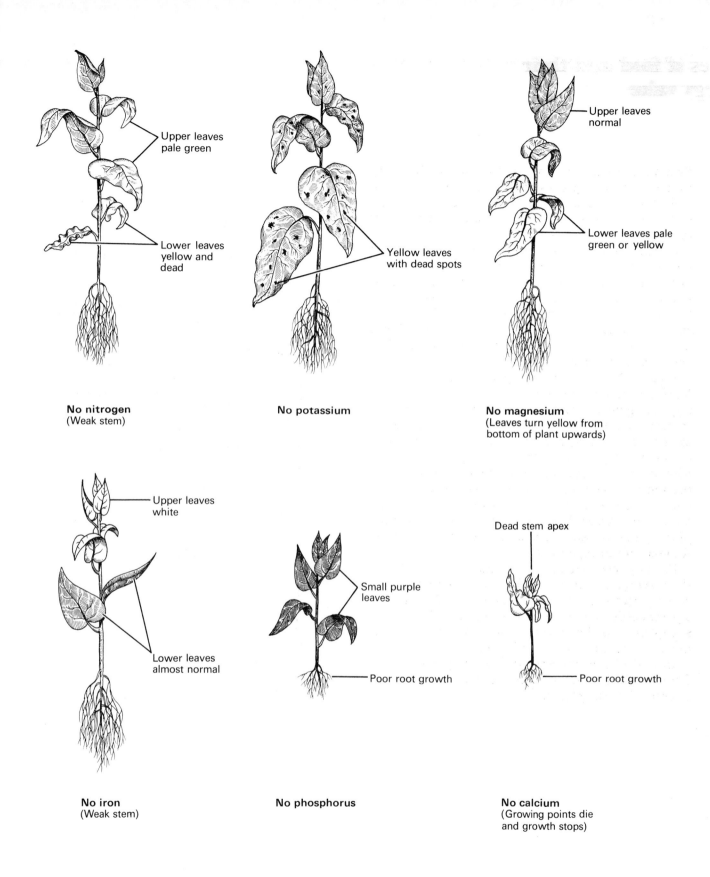

Fig. 19.2 Symptoms of some mineral deficiency diseases

Types of food and their energy value

Carbohydrates

Carbohydrates are sugary and starchy foods. These foods form the body's main source of energy. But if too much carbohydrate is eaten the body changes it into fat and stores it under the skin and around various body organs. Sugary foods include sweet fruits (strawberries, grapes, peaches, oranges, melons, cherries, gooseberries, etc.) cane and beet sugar, glucose, jam, treacle, honey, toffee, and chocolate. Starchy foods include bread, potatoes, rice, spaghetti, beans, peas, cakes, pastries, and buns.

Fats and oils

Fats and oils are the body's main stored foods. When they are eventually used to provide energy 1 gram of fat releases twice as much energy as 1 gram of carbohydrate or protein. The layer of fat under the skin insulates the body, keeping it warm in cold weather. Fatty foods include milk and cream, butter, margarine, lard, suet, dripping, cheese, and fatty meat such as bacon and pork. Oily foods include peanuts, olives, and the livers of fish such as cod and halibut.

Proteins

Proteins are used for growth, and for repairing damage to the body tissues, caused for example by cuts and grazes. Protein foods include cheese, eggs, meat (e.g. beef, pork, mutton), liver, fish, and milk.

Energy value of food

All foods contain stored energy, but the energy cannot be used by the body until it is released by the chemical reactions of respiration. The amount of energy which the body can obtain from food is now measured in units called **kilojoules** (abbreviated to kJ). For example, the body can obtain 17 kJ of energy from 1 gram of carbohydrate; 38 kJ of energy from 1 gram of fat or oil; and 17 kJ of energy from 1 gram of protein. Until recently, the amount of energy obtained from food was measured in **calories**, but these are now obsolete.

Fig. 20.1 Sort these foods into carbohydrates, fats, and proteins. Remember that some can be placed under more than one of these headings.

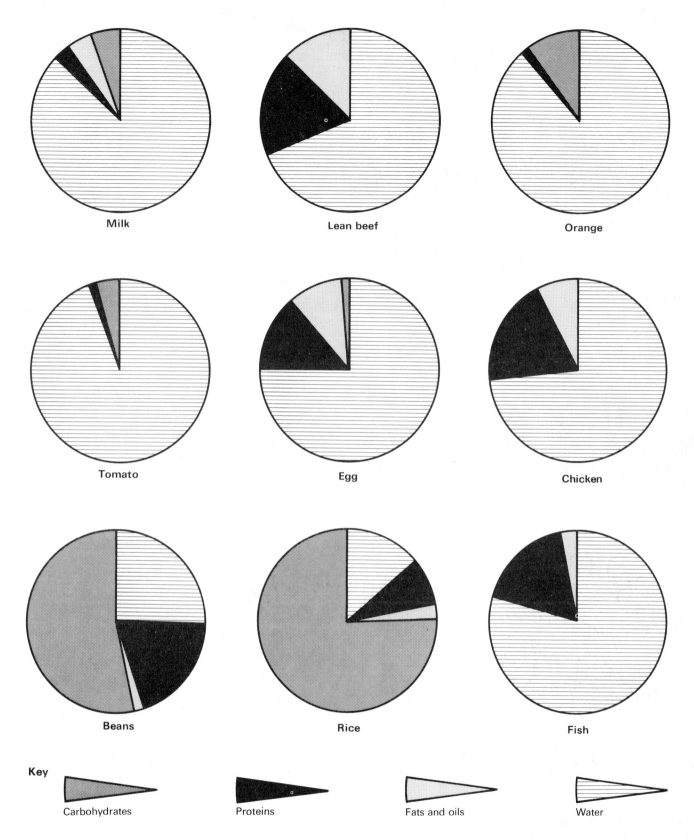

Key

Carbohydrates   Proteins   Fats and oils   Water

Fig. 20.2   Diagram showing the proportions of carbohydrates, proteins, fats, and water in some common foods
Humans require about fifteen different minerals in their diet. Most of these are supplied by meat, eggs, milk, green vegetables, and fruit

21

Vitamins and minerals

Vitamins are chemicals which are essential for growth and general health. The body needs only very small amounts of vitamins, but if they are missing from the diet illnesses called **vitamin deficiency diseases** develop.

Vitamin A keeps the skin and bones healthy, helps prevent infection of the nose and throat, and is necessary for vision in dim light. Lack of vitamin A causes poor night vision, and increases the chances of infection of the nose and throat. Vitamin A is found in carrots, milk, fish-liver oils, and green vegetables.

Vitamin B$_1$ helps the body obtain energy from food. Lack of it reduces growth, and causes **beri-beri**, a disease in which the limbs are paralysed. Vitamin B$_1$ is found in yeast, wholemeal bread, nuts, peas, and beans.

Vitamin B$_2$ enables the body to obtain energy from food. Lack of it causes stunted growth, cracks in the skin around the mouth, an inflamed tongue, and damage to the cornea of the eye. Vitamin B$_2$ is found in liver, milk, eggs, yeast, cheese, and green vegetables.

Vitamin B$_{12}$ enables the body to form protein and fat, and to store carbohydrate. Lack of it causes **pernicious anaemia** (failure to produce haemoglobin for red blood cells). Vitamin B$_{12}$ is found in liver, meat, eggs, milk, and fish.

Vitamin C is destroyed by cooking, grating, or mincing food. It disappears from food if it is stored for long periods. Vitamin C helps wounds to heal, and is needed for healthy gums and teeth. Lack of it causes **scurvy**, a disease in which the gums become soft, the teeth grow loose, and wounds fail to heal properly. Vitamin C is found in oranges, lemons, black currants, green vegetables, tomatoes, and potatoes.

Vitamin D enables the body to absorb calcium and phosphorus from food. These chemicals are needed to make bones and teeth. Lack of vitamin D causes **rickets** (soft weak bones, which bend under pressure). Vitamin D is found in liver, butter, cheese, eggs, and fish.

Vitamin K is needed to make blood clot in wounds. Lack of it can cause **haemorrhage** (excessive bleeding) whenever the skin is broken. Vitamin K need not be obtained from food. It is made by bacteria which live inside the digestive system.

Mineral	Daily requirement mg (1 mg = 0.001 g)	Function in the body
Sodium chloride (common salt)	5–10	Blood plasma is almost 1% salt. Salt is also needed for digestion, and to enable impulses to pass along nerve fibres.
Potassium	2	Needed for muscular contraction.
Magnesium	0·3	
Phosphorus	1·5	Forms a large part of bones and teeth. Also needed for chemical reactions of respiration.
Calcium	0·8	Calcium salts form an important part of bones and teeth.
Iron	0·01	A large part of haemoglobin is made up of iron. Haemoglobin is the substance which gives blood its red colour. It transports oxygen to the tissues.
Copper	0·001	Enables the body to use iron. Also needed for growth.
Manganese	0·003	Needed for growth.
Iodine	0·00003	Needed by the thyroid gland, which produces hormones. A lack of iodine causes the disease goitre, in which the thyroid gland becomes enlarged.

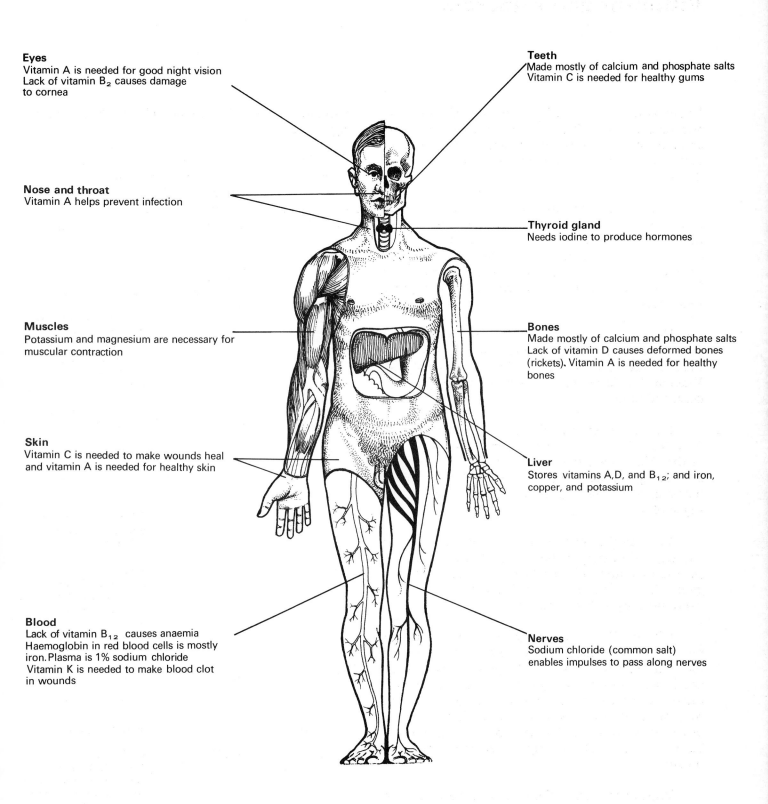

Eyes
Vitamin A is needed for good night vision
Lack of vitamin B_2 causes damage
to cornea

Nose and throat
Vitamin A helps prevent infection

Muscles
Potassium and magnesium are necessary for
muscular contraction

Skin
Vitamin C is needed to make wounds heal
and vitamin A is needed for healthy skin

Blood
Lack of vitamin B_{12} causes anaemia
Haemoglobin in red blood cells is mostly
iron. Plasma is 1% sodium chloride
Vitamin K is needed to make blood clot
in wounds

Teeth
Made mostly of calcium and phosphate salts
Vitamin C is needed for healthy gums

Thyroid gland
Needs iodine to produce hormones

Bones
Made mostly of calcium and phosphate salts
Lack of vitamin D causes deformed bones
(rickets). Vitamin A is needed for healthy
bones

Liver
Stores vitamins A, D, and B_{12}; and iron,
copper, and potassium

Nerves
Sodium chloride (common salt)
enables impulses to pass along nerves

Fig. 21.1 Functions of the main vitamins and minerals
required in humans

Digestion, absorption, and assimilation

Digestion is the process which makes food soluble. Food must be soluble before it can be absorbed into the body and used to provide energy, or for growth and repair of body tissues.

In humans food is digested inside a tube called the **alimentary canal**. This tube starts at the mouth and ends at the anus. In certain places the walls of this tube produce digestive juices containing chemicals called **digestive enzymes**. Digestive enzymes break food down into chemically simpler substances (one theory of how this happens is explained in Figure 22.1).

Digestive enzymes are divided into groups according to the type of food they digest. **Amylases** are a group of enzymes which digest starchy foods into sugars such as glucose. **Lipases** are enzymes which digest fats and oils into simpler substances called fatty acids and glycerol. **Proteases** digest proteins into simpler substances called amino acids.

When digestion is complete, **absorption** takes place. Absorption is the passage of digested (soluble) food through the wall of the alimentary canal into the blood-stream. The food is then transported to all the cells of the body, where a process called **assimilation** occurs.

Assimilation is the name for the processes by which cells use simple, soluble foods as raw materials for building up the complex substances of which the body is made.

Most foods contain substances which cannot be digested. Humans, for instance, have no enzymes which can digest the cellulose cell walls in foods taken from plants. Cellulose and other indigestible materials are called **faecal matter**, or **faeces**. Faeces pass out of the body through the anus. This process is called **defecation**.

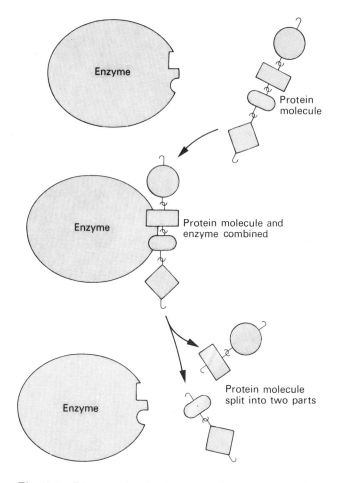

Fig. 22.1 Diagram showing how digestive enzymes work. Protein molecules are chains of linked parts called amino acids. Digestive enzymes break up these chains into small pieces and eventually into single amino acids. The diagram shows how two particular amino acids touch a special part of an enzyme molecule and while they are in this position undergo a chemical change. This breaks the link between the amino acids, and splits the protein molecule into two smaller pieces

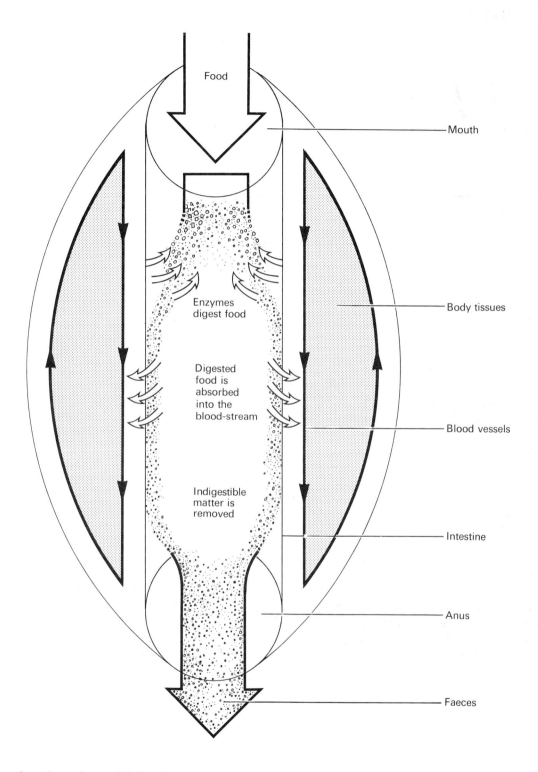

Food

Mouth

Enzymes
digest food

Body tissues

Digested
food is
absorbed
into the
blood-stream

Blood vessels

Indigestible
matter is
removed

Intestine

Anus

Faeces

Fig. 22.2 Diagram of digestion, absorption, and defecation.
This diagram shows the gut as a straight tube from mouth to
anus. Enzymes are released into this tube, food is broken down
into soluble form and absorbed into the blood-stream, and
indigestible matter is passed out of the anus

23

Teeth

Teeth grow in holes called sockets in the jaw bones. Figure 23.1 shows the parts of a tooth, and how teeth are held in place.

There are four types of tooth: **incisors** are at the front of the mouth, and **canines**, **premolars**, and **molars** are at the sides. The type of food that an animal can eat depends on the shape of its teeth.

Omnivore teeth

Omnivores can eat both plant and animal foods. Humans, for example, have chisel-shaped incisors (Fig. 23.2B). These are used to bite lumps off a piece of food. The canines are longer and more pointed and also help in biting. The crowns of premolar and molar teeth have projections called **cusps** which crush food and grind it into small pieces. This process is called chewing.

Carnivore teeth

Carnivores, such as the cat and dog families, are meat eaters. Their teeth help them catch, kill, and eat other animals, especially herbivores. Dogs, for example (Fig. 23.2C), have small incisors which they use to clean fur and cut flesh close to the bone. They grasp their prey with their long, dagger-like canines to prevent it escaping. The canines often kill the prey. The massive **carnassial** teeth (Fig. 23.2D) are used to crack bones and cut flesh into pieces small enough to swallow.

Herbivore teeth

Herbivores, such as sheep and cattle, have teeth that help them eat large quantities of plant food, especially grass.

In sheep, for example (Fig. 23.2E), there are no upper incisors. In the position where they would have occurred in other animals there is a pad of hard skin. When a sheep is eating the lower incisors move

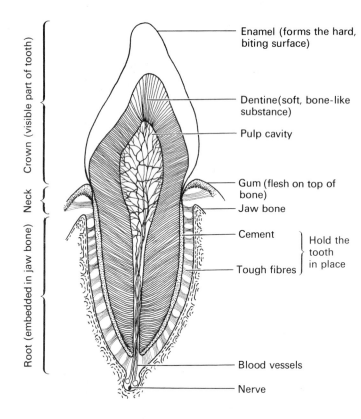

Fig. 23.1   Structure of a tooth

sideways across this pad. This movement cuts off grass blades close to the roots. Sheep have no canine teeth. Sheep premolars and molars wear down gradually, forming hard, sharp ridges of enamel; but the teeth never wear away completely because they grow throughout the animal's life. When a sheep chews the jaws move sideways and pull the ridged teeth across each other (Fig. 23.2F). This grinds the grass into a pulp.

Fig. 23.2 (right)   Teeth of omnivores, carnivores, and herbivores

46

A Omnivore (human)

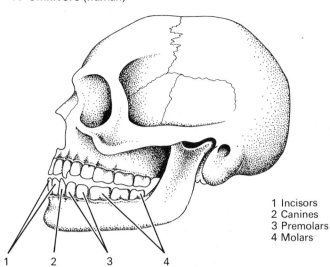

1 Incisors
2 Canines
3 Premolars
4 Molars

1 2 3 4

B Types of human teeth

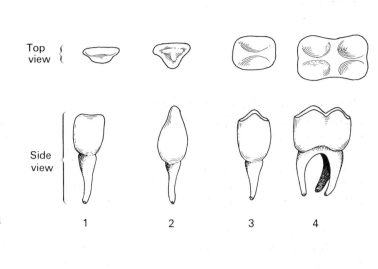

Top view {

Side view {

1 2 3 4

C Carnivore (dog)

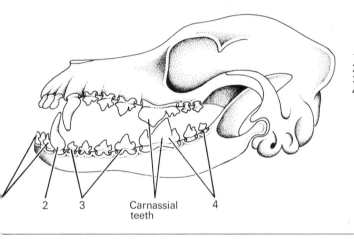

1 Incisors
2 Canines
3 Premolars
4 Molars

2 3 Carnassial teeth 4

D Carnassial teeth

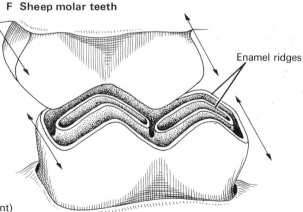

Front view   Side view

Note how carnassial teeth overlap when the mouth is closed

E Herbivore (sheep)

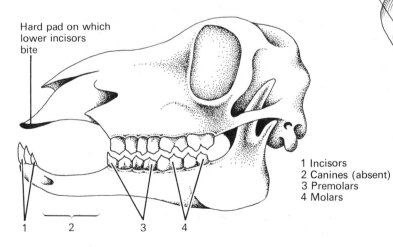

Hard pad on which lower incisors bite

1 Incisors
2 Canines (absent)
3 Premolars
4 Molars

1 2 3 4

F Sheep molar teeth

Enamel ridges

The enamel ridges on the top and bottom teeth slide over each other as the jaws move from side to side

47

24

Human digestive system

Digestion is the process which makes food soluble. Soluble food can be absorbed into the blood-stream and transported to all parts of the body. Food is digested by **enzymes**: chemicals contained in juices produced by the digestive system.

Digestion in the mouth

Chewing mixes food with **saliva**, produced by the **salivary glands**. Saliva moistens food so that it passes easily into the **oesophagus** when it is swallowed (Fig. 24.1). Muscles in the oesophagus wall contract, pushing food from the throat to the stomach. Similar muscles in the wall of the gut push food through the rest of the digestive system. These muscular contractions are called **peristalsis** (Fig. 24.2).

Digestion in the stomach

The stomach has rings of muscle at both ends called **sphincters** (Fig. 24.3A). After a meal both sphincters contract and keep food in the stomach for about an hour. Muscles in the stomach wall contract rhythmically, mixing food with a liquid called **gastric juice**. This contains **pepsin**, an enzyme which digests protein, and a weak solution of hydro-chloric acid, which kills germs and helps the enzyme to work. When the lower sphincter opens, partly digested food moves into the **duodenum**, which is the first part of the small intestine (Fig. 24.3A and B).

Digestion in the small intestine

In the duodenum food is mixed with **bile**. Bile is made in the **liver**, and stored in the **gall bladder**. It is carried to the duodenum through the **bile duct**. Bile helps digestion by changing fatty foods into an **emulsion**, which means the fats and oils are broken down into minute droplets. Bile also neutralizes stomach acid. This enables enzymes in the small intestine to work. Food is mixed with enzymes from the **pancreas**, and later with more enzymes from the **ileum** wall. These enzymes digest food completely. When it is soluble it is absorbed into the blood-stream.

Functions of the large intestine

The substances in food which cannot be digested pass into the large intestine (Fig. 24.3B). Here, water and salt are removed. The remaining **faecal matter**, or **faeces**, pass out of the body at intervals through the anus.

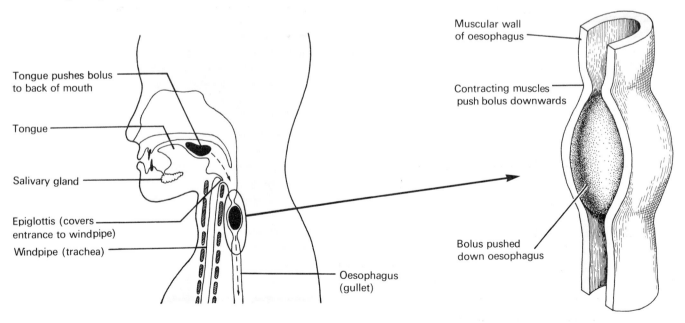

Fig. 24.1 Swallowing

Fig. 24.2 Oesophagus cut open to show peristalsis

A Stomach and duodenum

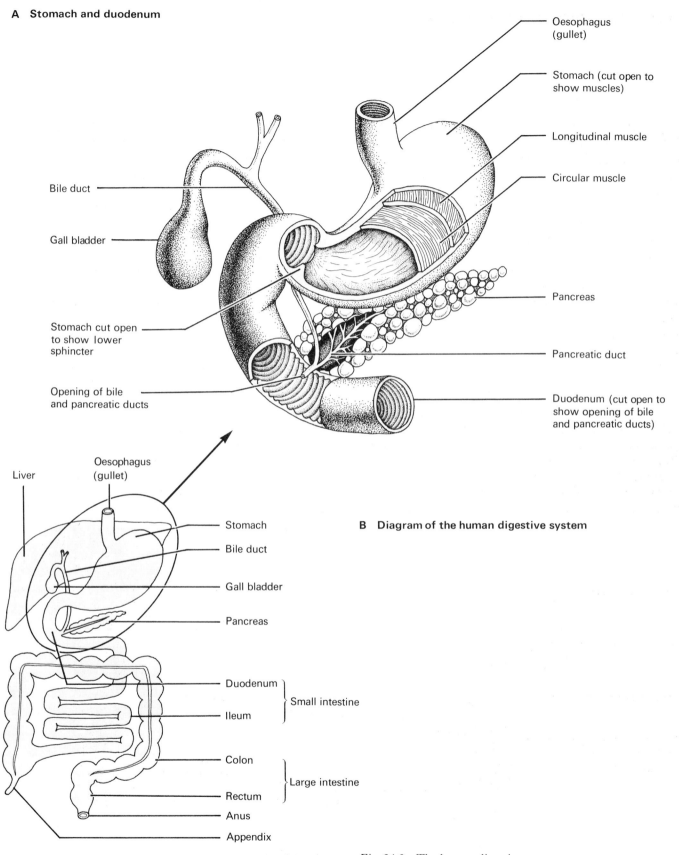

Oesophagus
(gullet)

Stomach (cut open to
show muscles)

Longitudinal muscle

Circular muscle

Bile duct

Gall bladder

Pancreas

Pancreatic duct

Stomach cut open
to show lower
sphincter

Opening of bile
and pancreatic ducts

Duodenum (cut open to
show opening of bile
and pancreatic ducts)

Liver

Oesophagus
(gullet)

B Diagram of the human digestive system

Stomach

Bile duct

Gall bladder

Pancreas

Duodenum

Ileum

Small intestine

Colon

Large intestine

Rectum

Anus

Appendix

Fig. 24.3   The human digestive system

49

Absorption of digested food

Absorption is the movement of digested food through the intestine wall into the blood-stream. From here, food is transported to all parts of the body. Before food can be absorbed it must be completely broken down into soluble substances: carbohydrates are broken down into sugar (glucose); fats and oils into fatty acids and glycerol; and proteins into amino acids. Vitamins and minerals are not digested: they are already soluble.

A small amount of food is absorbed in the stomach, but absorption takes place mainly in the **duodenum** and **ileum**, the two parts of the small intestine (Fig. 24.3B). The whole of the small intestine is lined with tiny finger-like projections called **villi** (*singular:* **villus**, Fig. 25.2). Each villus is about 1 mm long, and there are roughly forty per square millimetre in the ileum wall. Altogether there are about five million villi in the small intestine. The villi give the intestine a far greater surface area through which food can be absorbed than if it had a smooth lining.

In each villus there is a network of capillary blood vessels, and a single central vessel filled with a liquid called **lymph** (Fig. 25.2C). Digested food, and foods which can be absorbed directly into the blood-stream (vitamins, minerals, glucose sugar, fatty acids, glycerol, and amino acids), pass through the walls of the villi into the blood capillaries. Capillaries from all the villi drain into a larger blood vessel called the **hepatic portal vein**. This vessel carries food to the liver where it is stored, processed in various ways, and released as the body requires it for all its activities.

Undigested droplets of oil pass through the surface of each villus into the central lymph vessel. From here they are carried to larger lymph vessels which drain into the blood-stream.

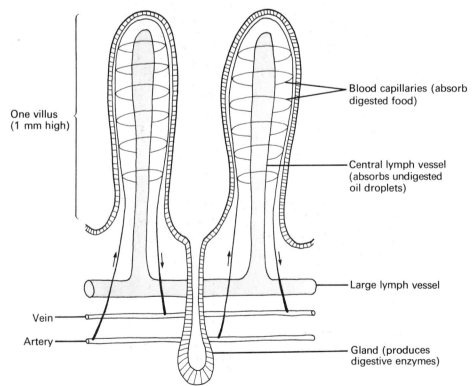

One villus
(1 mm high)

Blood capillaries (absorb digested food)

Central lymph vessel (absorbs undigested oil droplets)

Large lymph vessel

Vein

Artery

Gland (produces digestive enzymes)

Fig. 25.1   Diagram of the ileum wall

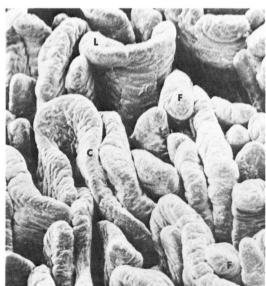

A variety of shapes of villi from the ileum. F is finger-like; L is leaf-like; and C is coiled into a loop

A Ileum cut open to show villi

Fig. 25.2   Structure of the ileum

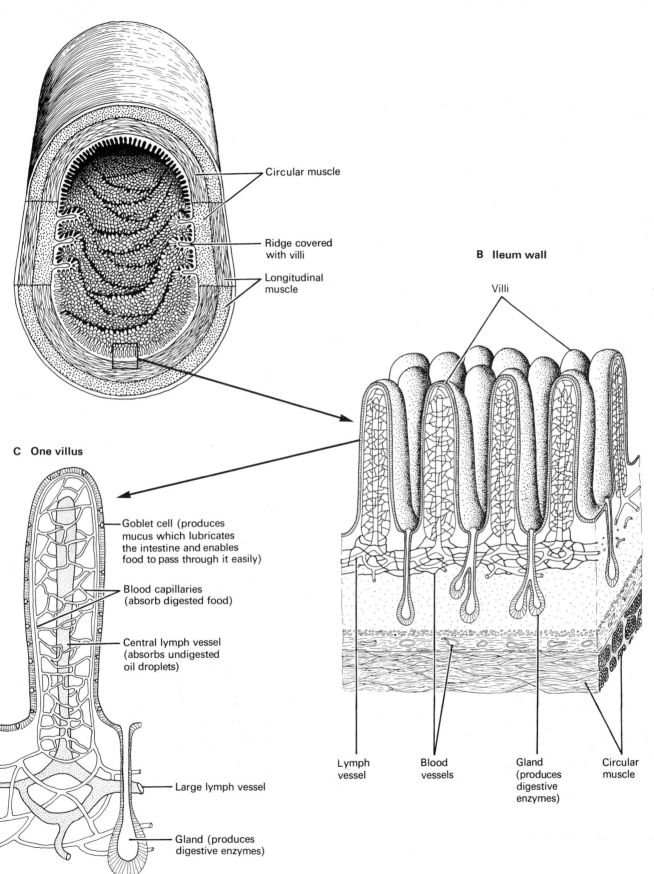

Circular muscle

Ridge covered
with villi

Longitudinal
muscle

B Ileum wall

Villi

C One villus

Goblet cell (produces
mucus which lubricates
the intestine and enables
food to pass through it easily)

Blood capillaries
(absorb digested food)

Central lymph vessel
(absorbs undigested
oil droplets)

Large lymph vessel

Gland (produces
digestive enzymes)

Lymph
vessel

Blood
vessels

Gland
(produces
digestive
enzymes)

Circular
muscle

Feeding in insects

The mouth of an insect is surrounded by various structures called **mouth parts**. The shape of an insect's mouth parts enables it to eat certain types of food. Locusts and cockroaches have biting and chewing mouth parts, and can therefore eat solid foods. Houseflies, butterflies, and mosquitoes have tubular mouth parts with which they suck up liquid foods.

Locust mouth parts

Locusts eat plant material, especially grasses. In certain parts of Africa swarms of locusts often eat every scrap of vegetation over huge areas. Locust mouth parts have the following features (Fig. 26.1).

Labrum   At the front of the mouth is a large flap called the labrum, or upper lip. Its function is to taste food, in the same way as the human tongue.

Mandibles   The mandibles, or jaws, have powerful muscles and sharp saw-like teeth. The teeth cut up food into small pieces.

Maxillae   The maxillae also have teeth, which help to cut up food. Sense organs called **palps** are attached to the maxillae. These are used to detect food.

Labium   The labium, or lower lip, has no jaws, but does have a pair of palps which act as sense organs.

Housefly mouth parts

A housefly has no mandibles (jaws). It sucks up liquid foods through a long tube called a **proboscis** (Fig. 26.2).

The proboscis contains a food channel and a salivary channel. Both these channels open into many tiny grooves on the under-surface of two flaps at the base of the proboscis. When a housefly feeds, saliva containing digestive juices is squirted on to the food through the salivary channel. The saliva partially digests the food, turning it into a liquid. This liquid is then sucked up the food channel into the digestive system of the insect. It is then digested completely and absorbed.

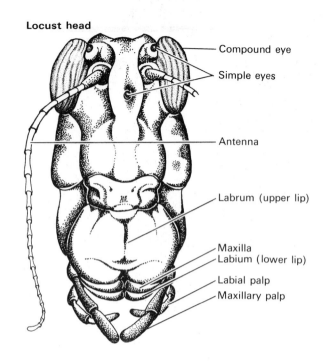

Locust head

- Compound eye
- Simple eyes
- Antenna
- Labrum (upper lip)
- Maxilla
- Labium (lower lip)
- Labial palp
- Maxillary palp

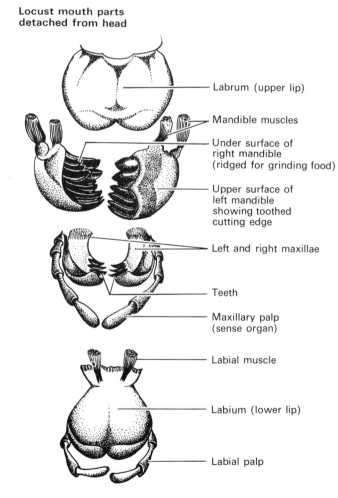

Locust mouth parts detached from head

- Labrum (upper lip)
- Mandible muscles
- Under surface of right mandible (ridged for grinding food)
- Upper surface of left mandible showing toothed cutting edge
- Left and right maxillae
- Teeth
- Maxillary palp (sense organ)
- Labial muscle
- Labium (lower lip)
- Labial palp

Fig. 26.1   Locust head and mouth parts

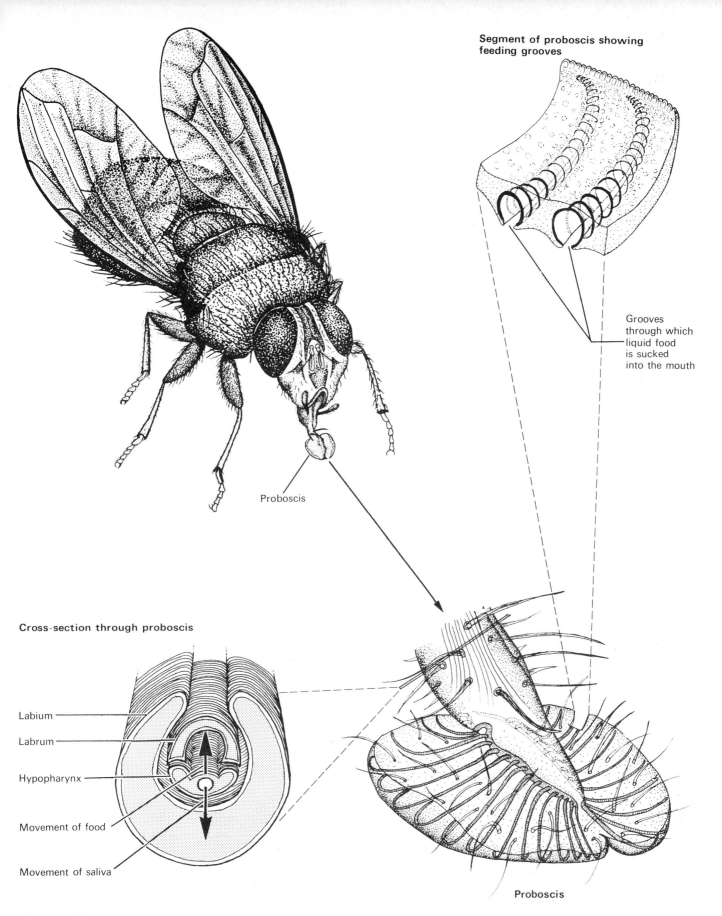

Segment of proboscis showing
feeding grooves

Grooves
through which
liquid food
is sucked
into the mouth

Proboscis

Cross-section through proboscis

Labium

Labrum

Hypopharynx

Movement of food

Movement of saliva

Proboscis

Fig. 26.2 Housefly mouth parts

Saprophytes *(Mucor)*

Many fungi and bacteria live as **saprophytes**. Saprophytes feed on dead organisms. They produce digestive juices which dissolve dead organisms, causing them to decay. The digested substances are then absorbed into the saprophyte's body.

Mucor, sometimes called pin mould, is a fungal saprophyte. Like most fungi, *Mucor* consists of many fine hollow threads called **hyphae** (Fig. 27.2). The hyphae in a mould colony are collectively called the **mycelium**. Special **absorptive hyphae** grow down into the food, which is then digested and absorbed. Hyphae on the surface of the colony carry out asexual and sexual reproduction.

Mucor spreads from place to place by producing billions of microscopic spores. This is done by asexual reproduction (Fig. 27.1). The tiny oval spores, containing many nuclei, form within swellings which develop at the tips of upright hyphae. These swellings called **sporangia** (*singular :* **sporangium**) eventually burst open and expose the ripe spores. In some types of *Mucor* the spores are blown away by the wind. In other types the spores are covered with sticky jelly, and are carried away on the feet and bodies of flies which feed on the decaying food on which the mould grows. If a spore reaches a substance on which it can grow it bursts open and produces a new mycelium.

On rare occasions *Mucor* reproduces sexually by a process called **conjugation** (Fig. 27.3). Conjugation can only occur between hyphae of different colonies. There is no visible difference between the conjugating hyphae, but the two types are referred to as 'plus' and 'minus' strains. During conjugation a 'plus' and a 'minus' hypha come into contact and each forms a swelling full of nuclei. A wall forms behind each swelling, separating it from the rest of the mycelium. Then the wall between the two swellings dissolves, and a single compartment is formed. The nuclei from each swelling now fuse together in pairs : each 'plus' nucleus fusing with a 'minus' nucleus. Finally, a thick outer wall develops and the structure becomes a **zygospore**. When conditions are suitable for growth the zygospore bursts open and produces a single upright hypha. This hypha grows a sporangium and releases asexual spores. These may develop a new mycelium in favourable conditions.

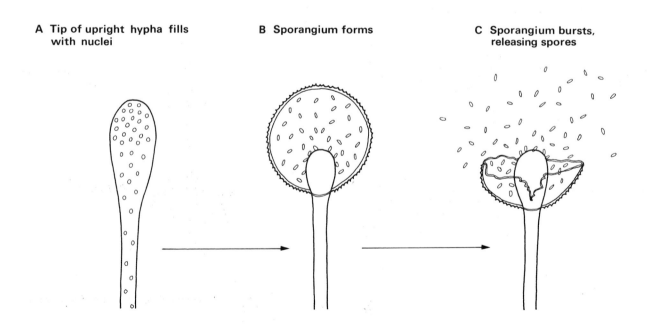

A **Tip of upright hypha fills with nuclei**

B **Sporangium forms**

C **Sporangium bursts, releasing spores**

Fig. 27.1 Diagram of asexual reproduction in *Mucor*

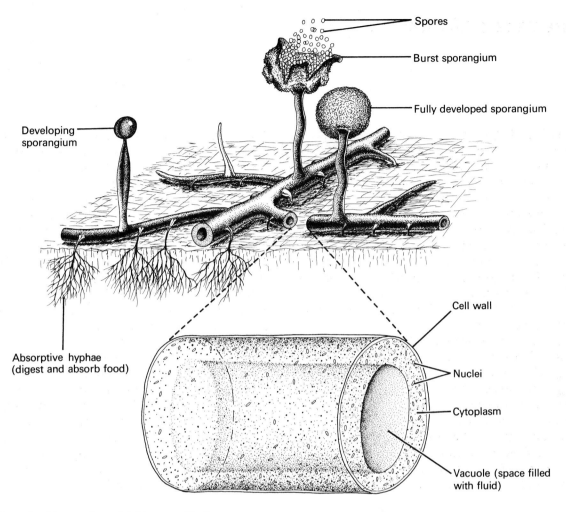

Spores

Burst sporangium

Fully developed sporangium

Developing sporangium

Cell wall

Nuclei

Cytoplasm

Vacuole (space filled
with fluid)

Absorptive hyphae
(digest and absorb food)

Fig. 27.2   Part of a *Mucor* colony (highly magnified)

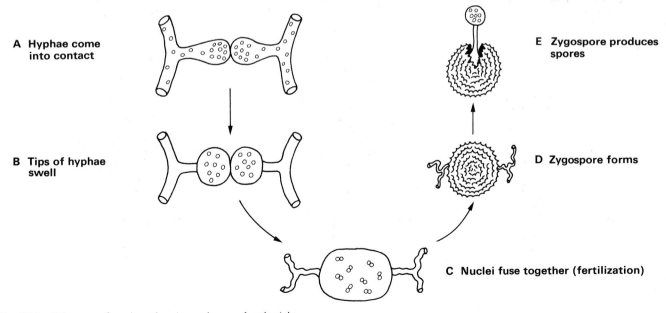

**A Hyphae come
into contact**

**B Tips of hyphae
swell**

C Nuclei fuse together (fertilization)

D Zygospore forms

**E Zygospore produces
spores**

Fig. 27.3   Diagram of conjugation (sexual reproduction) in
Mucor

Parasites (tapeworm)

A **parasite** is an organism which obtains its food from the living body of another organism called the **host**. Tapeworms are parasites with two hosts. The **primary host** is always a vertebrate animal, and carries the adult worm in its intestine. The **secondary host** carries a partly developed worm called a **bladderworm** in its tissues. If the secondary host is eaten by the primary host, the bladderworms develop into adult tapeworms in the primary host's intestine.

In *Taenia solium*, the pork tapeworm, the primary host is man. The adult worm consists of a head with suckers and hooks to attach itself to the intestine wall of its host (Figs. 28.1 and 28.2A). The flat 'tape' of the worm may be as much as 8 m long and may consist of as many as 1000 segments. The tape absorbs digested food from the host's intestine. It is not digested by its host because it is covered by an enzyme-resistant skin called the **tegument**.

At first each segment of the tape consists of both male and female sex organs (in other words it is hermaphrodite), but after fertilization these disappear and are replaced by up to 50 000 eggs. At this stage the segments drop off the end of the tape and pass out of the host's body with the faeces.

In places where there are poor sewage disposal arrangements, tapeworm eggs may be eaten by the secondary host, which is a pig. If this happens, the shell of the egg is dissolved by the pig's digestive juices, releasing an **embryo** 0.02 mm in diameter, armed with six hooks (Fig. 28.2B). The embryo uses these hooks to burrow into the pig's intestine until it reaches a blood vessel. It is then carried around the body. Embryos which are carried into muscle tissue develop into bladderworms. These are bags of liquid about 7 mm long containing a tapeworm head, inside out. Bladderworms do not harm the pig, and develop no further unless the muscle (pork meat) which contains them is eaten by humans. If pork containing bladderworms is not cooked long enough to kill them they unfold in the human intestine so that their hooks and suckers are on the outside. The head now becomes attached to the intestine and a tape develops behind it.

Infection by tapeworms can be prevented by adequate sewage disposal arrangements which ensure that the worm eggs cannot enter pig food and so complete their life cycle. The worms will then gradually die out. Imported meat should be carefully inspected for bladderworms, which can be seen clearly with the naked eye.

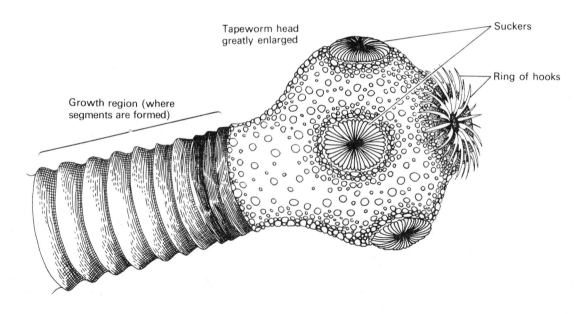

Tapeworm head
greatly enlarged

Suckers

Ring of hooks

Growth region (where
segments are formed)

Fig. 28.1   Head and growth region of a tapeworm

A Inside primary host (man)

B Inside secondary host (pig)

Tapeworm head released when poorly cooked pork containing bladderworms is eaten by primary host

Bladderworm

Tapeworm head embedded in intestine wall

Bladderworms embedded in muscle

Intestine

Blood vessel

Embryos transported by bloodstream to muscles

Eggs containing an embryo with six hooks

Embryo burrows through pig's intestine wall

Ripe segments full of eggs break off tape and pass out of host in its faeces

Embryo

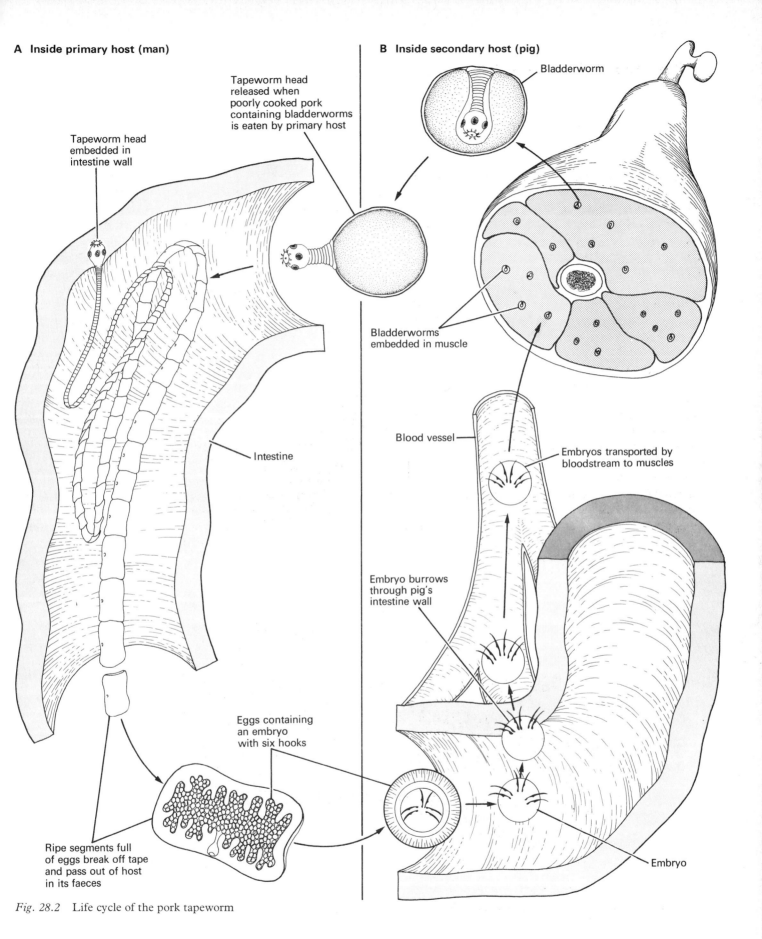

Fig. 28.2   Life cycle of the pork tapeworm

57

Symbiosis and commensalism

Many kinds of organisms live in groups of the same species. For example, elephants live in herds, and birds live in flocks. But sometimes two entirely different types of organism live together, and co-operate with each other in such a way that they both benefit from the partnership. This type of relationship is called **symbiosis**.

A well known example of symbiosis is the relationship between nectar-seeking insects and certain flowering plants. Bees, for example, benefit by obtaining nectar and pollen from flowers; while the plants benefit from being cross-pollinated as bees transfer pollen from one flower to another (Fig. 29.1).

A hermit crab has a long curved abdomen which it inserts into the coiled interior of an empty whelk shell (Fig. 29.4) for protection. The crab gains extra protection by finding a particular type of sea anemone which it places on top of its shell with its pincers. The crab is protected from predators, which are kept at a distance by the stinging cells on the anemone's tentacles. The anemone benefits by obtaining scraps of food discarded by the crab.

A much closer symbiotic relationship exists in plant-like organisms called lichens (Fig. 29.2). A lichen is made up of green algae enclosed within a dense mass of cells and threads produced by a fungus. The fungus in a lichen obtains carbohydrates and oxygen which are produced as the algae carry out photosynthesis. The algae obtain water and minerals from the fungus, and are also sheltered and protected from drying up. Lichens are so hardy that they can live attached to bare rocks high on mountain sides.

Commensalism is a type of relationship between different organisms in which only *one* partner derives benefit. Sharks, for example, often have sucker fish attached to them and pilot fish swimming nearby (Fig. 29.3). The shark neither suffers harm nor benefits from their presence, but the sucker fish obtains 'free' transport, and it shares discarded food from the shark's meals with the pilot fish. Both the sucker fish and the pilot fish are protected from predators by the presence of the shark.

It can be difficult to decide whether a relationship between organisms is symbiotic or commensal.

A Two types of lichen (actual size)

B Cross-section of a lichen

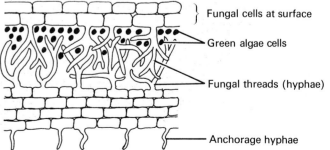

Fungal cells at surface

Green algae cells

Fungal threads (hyphae)

Anchorage hyphae

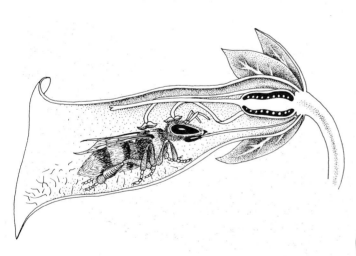

Fig. 29.1　A bee taking nectar from a flower

Fig. 29.2　Lichens

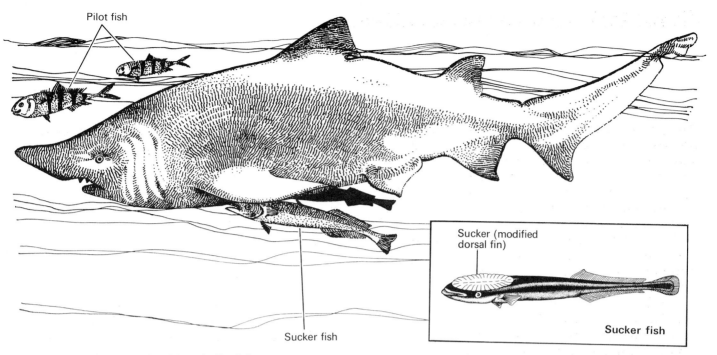

Fig. 29.3   Shark with sucker fish and pilot fish

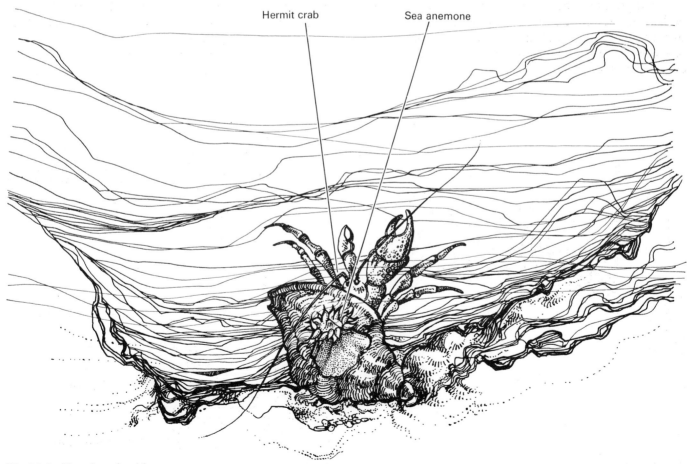

Fig. 29.4   Hermit crab with sea anemone

The transport systems of a plant

Substances are transported from one part of a plant to another through extremely narrow tubes in the root, stem, and leaves. Tubes called **xylem vessels** transport water and dissolved minerals from the roots to the leaves. Tubes called **phloem sieve tubes** transport sugar made by photosynthesis from the leaves to growing areas of the plant, and to food storage areas.

The walls of xylem vessels are strengthened by a substance called **lignin**, which occurs in the form of coils, rings, or thick layers perforated by tiny holes (Fig. 30.2B). Xylem vessels, and the thick pointed **xylem fibres** (also made of lignin) alongside them, form the woody 'skeleton' which supports a plant.

Phloem sieve tubes are divided into sections by sieve-like plates. Sieve tubes always have **companion cells** next to them, which presumably help them carry out their transport functions, although it is not yet clear exactly how they do this.

The transport system in a root

Plant roots take in water and dissolved minerals from the soil through **root hairs** (Fig. 30.1A). These hairs grow out of a band of cells close to the root tips (Fig. 30.1B). Water moves towards the centre of the root from the root hairs until it reaches an X-shaped mass of xylem vessels. Water then moves up these vessels to the stem. Food (sugar) moves down the root to the growing area at the root tips through phloem sieve tubes situated between the X-shaped areas of xylem.

The transport system in a stem

In the stems of most flowering plants xylem and phloem are arranged together in strands called **vascular bundles** (Fig. 30.2A). Xylem is usually situated nearest the centre of the stem, and is separated from phloem by a group of cells called the **cambium**. Cambium cells divide and grow into new xylem and phloem.

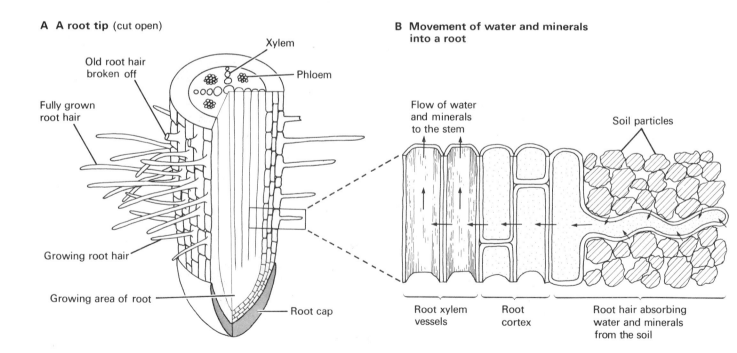

A A root tip (cut open)

Xylem

Old root hair broken off

Phloem

Fully grown root hair

Growing root hair

Growing area of root

Root cap

B Movement of water and minerals into a root

Flow of water and minerals to the stem

Soil particles

Root xylem vessels

Root cortex

Root hair absorbing water and minerals from the soil

Fig. 30.1 Structure of a root

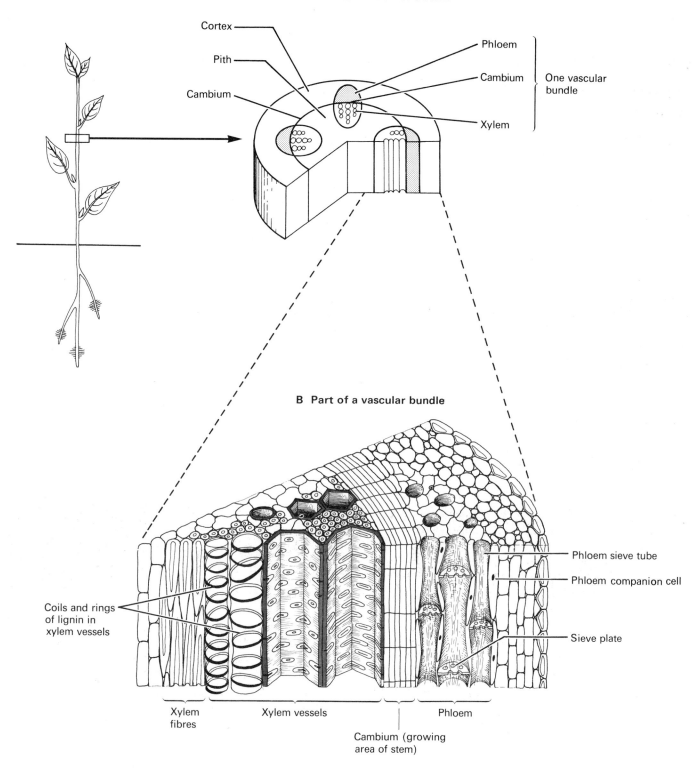

A Cross-section of a stem

Cortex

Pith

Cambium

Phloem

Cambium

Xylem

One vascular bundle

B Part of a vascular bundle

Coils and rings of lignin in xylem vessels

Phloem sieve tube

Phloem companion cell

Sieve plate

Xylem fibres

Xylem vessels

Cambium (growing area of stem)

Phloem

Fig. 30.2 Structure of a stem

Osmosis and transpiration

Plant roots take in water (but not minerals) from the soil by **osmosis**. Osmosis is a special kind of diffusion and is defined as follows: the diffusion of water through a semi-permeable membrane from a *weak* to a *strong* solution (Fig. 31.1).

Root hairs take in water by osmosis because their cell membranes are semi-permeable, and because their cell sap is a stronger solution than soil water. But as root hairs take up water their cell sap is diluted and soon becomes a weaker solution than the sap of cells deeper inside the root. Because of this the deeper cells take in water by osmosis from the root hairs. Water entering root hairs by osmosis gradually flows further into the root from cell to cell and eventually reaches the root xylem vessels (Fig. 31.3). Water flows into root xylem vessels with sufficient force to push it up into the stem xylem. This force is called **root pressure**. But root pressure is not enough to force water all the way up to the leaves. The energy for this comes from **transpiration**.

Transpiration is the loss of water from a plant by evaporation. Water evaporates mainly from the spongy mesophyll cells of a leaf into the spaces between these cells, and then escapes into the air through pores called **stomata** (Figs. 31.2 and 31.3). As mesophyll cells lose water by evaporation their sap becomes a stronger solution and they absorb water by osmosis from cells deeper in the leaf with weaker cell sap. The sap in these cells becomes stronger and absorbs water by osmosis from xylem vessels in the leaf veins.

Water removed from leaf xylem is replaced by water drawn up from stem xylem, and this is replaced by water forced up by root pressure. The result is a continuous flow of water (and dissolved minerals absorbed from the soil by another mechanism) from the roots to the leaves, where it is used in photosynthesis. This flow is called the **transpiration stream**. Transpiration is usually faster on warm sunny days than on cold dull days.

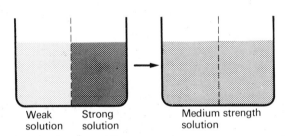

Weak solution   Strong solution   Medium strength solution

A Diffusion If a weak sugar solution is separated from a strong sugar solution by a membrane which allows both sugar and water to pass through it, sugar will diffuse from the strong to the weak solution and water will diffuse from the weak to the strong solution until both solutions are of equal strength

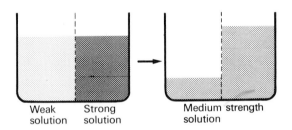

Weak solution   Strong solution   Medium strength solution

B Osmosis If a weak sugar solution is separated from a strong sugar solution by a membrane which allows *only* water to pass through it (a semi-permeable membrane), water will diffuse from the weak to the strong solution until both solutions are of equal strength

Fig. 31.1 Diagram showing diffusion and osmosis

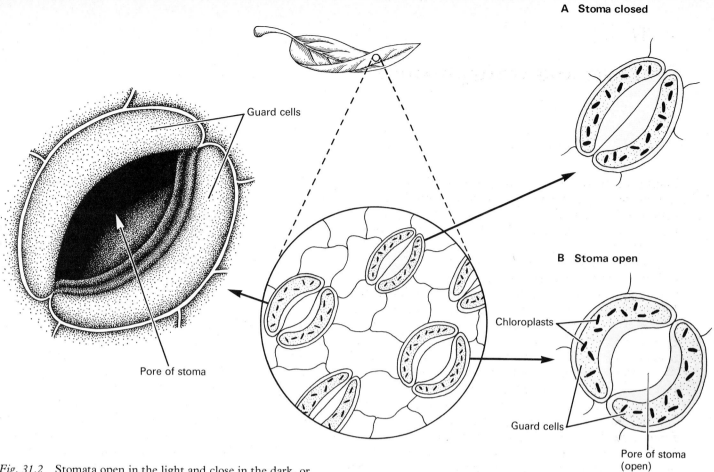

A **Stoma closed**

B **Stoma open**

Guard cells

Pore of stoma

Chloroplasts

Guard cells

Pore of stoma
(open)

Fig. 31.2 Stomata open in the light and close in the dark, or when a plant is short of water.

Fig. 31.3 Diagram showing how water flows through a plant (the transpiration stream)

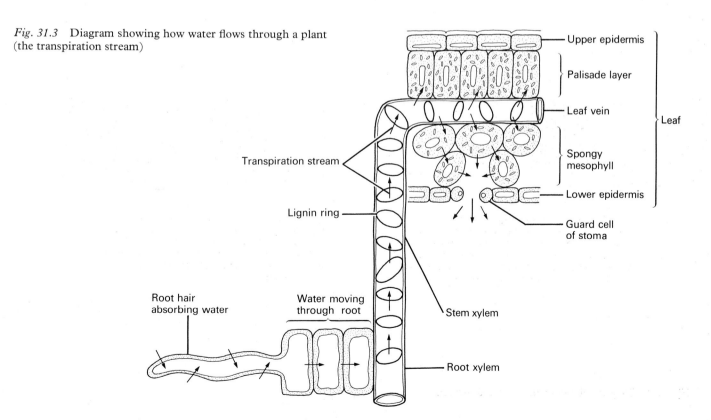

Upper epidermis

Palisade layer

Leaf vein

Leaf

Spongy mesophyll

Lower epidermis

Guard cell of stoma

Transpiration stream

Lignin ring

Stem xylem

Root hair absorbing water

Water moving through root

Root xylem

32

Blood and its functions

Blood consists of a liquid called **plasma**, which has **red cells**, **white cells**, and **platelets** floating in it.

Plasma

The word plasma is used to describe the liquid part of blood, that is, blood with the cells and platelets removed. Plasma is more than 90% water. The rest consists of dissolved substances such as food, waste products, hormones, and antibodies.

Red blood cells

Human red cells are disc-shaped and concave on both sides (Fig. 32.1A). Red cells have no nucleus. They live for only about four months, then they are destroyed by the spleen and some of their chemicals are re-used in the bone marrow to make new red cells. Red cells contain a chemical called **haemoglobin**, which gives them their red colour. Haemoglobin enables red cells to transport oxygen from the lungs to all parts of the body. As blood passes through vessels in the lungs haemoglobin picks up oxygen and changes into **oxyhaemoglobin**. As blood flows through the rest of the body oxy-

haemoglobin slowly releases oxygen to the body cells, and changes back into haemoglobin.

White blood cells

White cells are colourless, and all of them have a nucleus. About 75 per cent of white cells have a large, irregularly shaped nucleus (Fig. 32.1B). These white cells are called **phagocytes**, which literally means 'cell-eaters'. They can change shape like *Amoeba*, and destroy germs by engulfing and digesting them. To reach germs in wounds and other infected areas phagocytic white cells leave the blood-stream. They force their way through blood vessel walls and move through the liquid between body cells (Fig. 32.2D).

Platelets

Platelets are not complete cells. They are tiny fragments of cells made in the bone marrow (Fig. 32.1C). They help blood to clot in wounds. When blood clots it turns into a jelly which gradually hardens as it is exposed to the air. The clot prevents further bleeding and reduces the chances of infection.

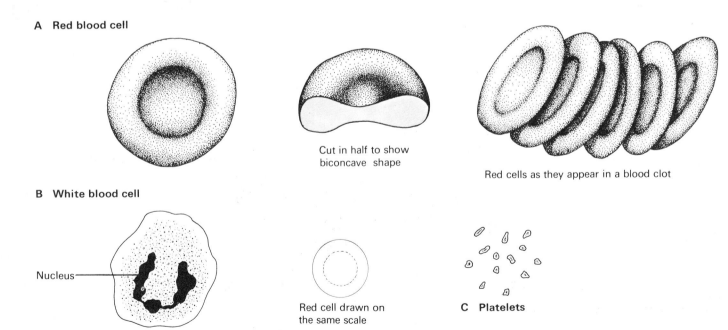

A Red blood cell

Cut in half to show biconcave shape

Red cells as they appear in a blood clot

B White blood cell

Nucleus

Red cell drawn on the same scale

C Platelets

Fig. 32.1 Blood cells

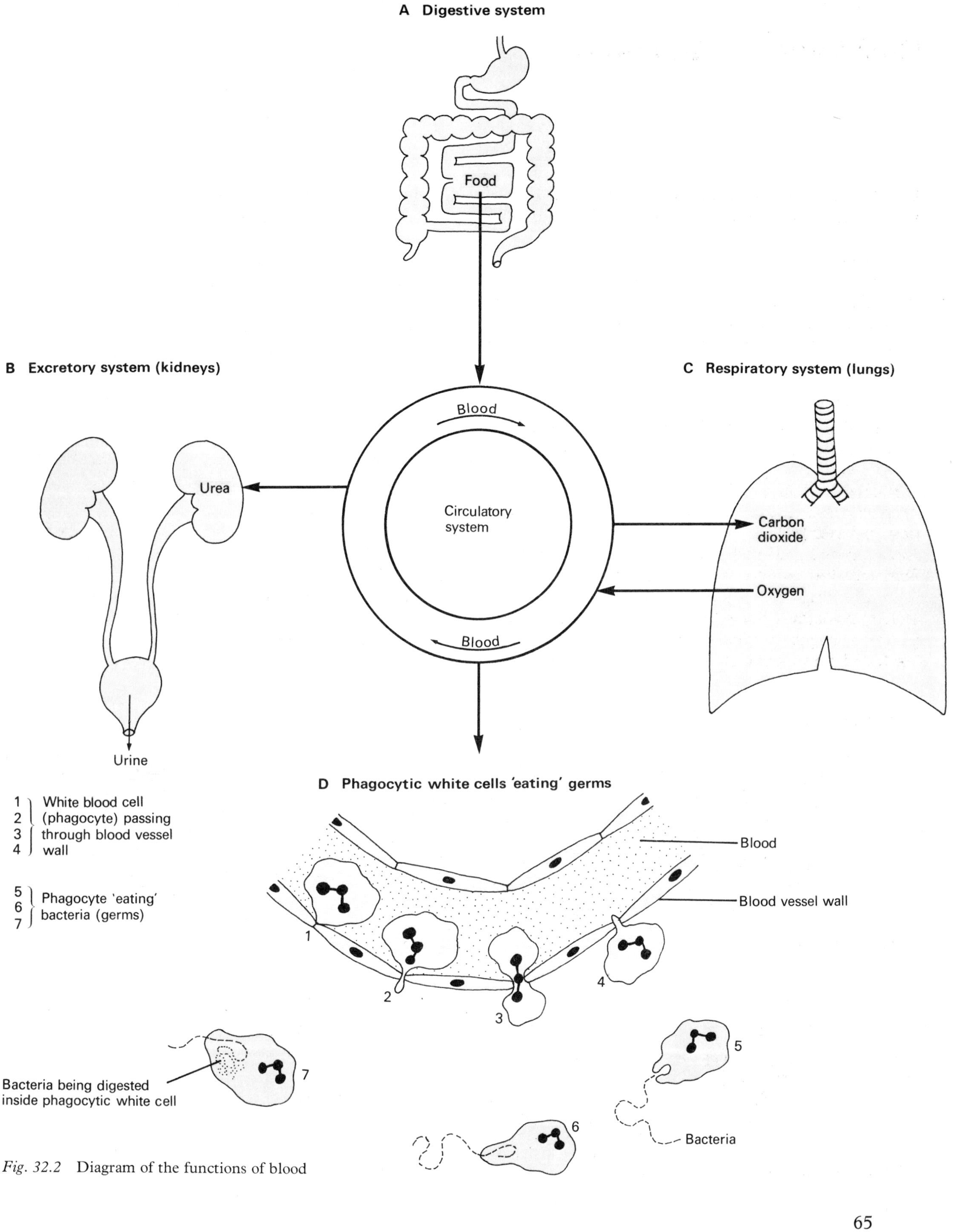

A Digestive system

Food

B Excretory system (kidneys)

Urea

Urine

C Respiratory system (lungs)

Blood

Circulatory
system

Blood

Carbon
dioxide

Oxygen

1 } White blood cell
2 } (phagocyte) passing
3 } through blood vessel
4 } wall

5 } Phagocyte 'eating'
6 } bacteria (germs)
7 }

D Phagocytic white cells 'eating' germs

Blood

Blood vessel wall

1

2

3

4

Bacteria being digested
inside phagocytic white cell

7

6

5

Bacteria

Fig. 32.2   Diagram of the functions of blood

65

The heart

The heart is a hollow bag with walls made of a special kind of muscle, called **cardiac muscle.** When cardiac muscle contracts it squeezes blood out of the heart into blood vessels which carry it to all parts of the body. When the cardiac muscle relaxes the heart fills with blood that is returning from its journey around the body. One contraction and relaxation of the heart is called a heart-beat. In humans, the heart normally beats from 60 to 70 times a minute. During exercise this may increase to about 150 times a minute.

The space inside the heart is divided into four compartments, called **chambers.** The two top chambers are called **atria,** or **auricles,** and their walls contain a thin layer of cardiac muscle. The two lower chambers are called **ventricles,** and their walls are made of a much thicker layer of cardiac muscle.

The heart contains a number of valves. These control the direction in which blood flows through the heart. The **bicuspid** and **tricuspid valves** ensure that blood flows only from the atria into the ventricles. The lower edges of these valves are held in place by tendons and muscles. **Semi-lunar valves** are situated at the points where blood flows out of the heart. The semi-lunar valves ensure that blood cannot flow back into the ventricles once it has left the heart.

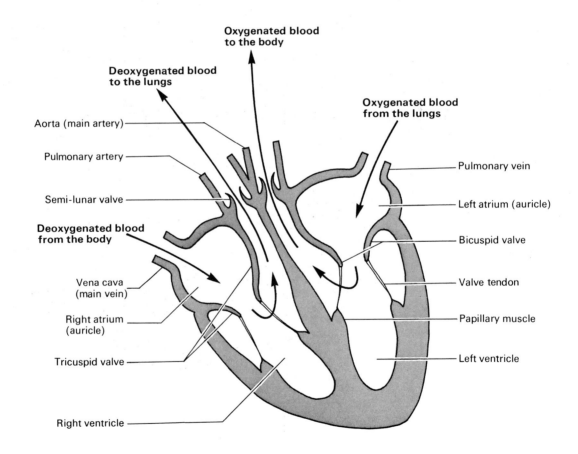

Fig. 33.1 Diagram of the heart, showing the direction of blood flow

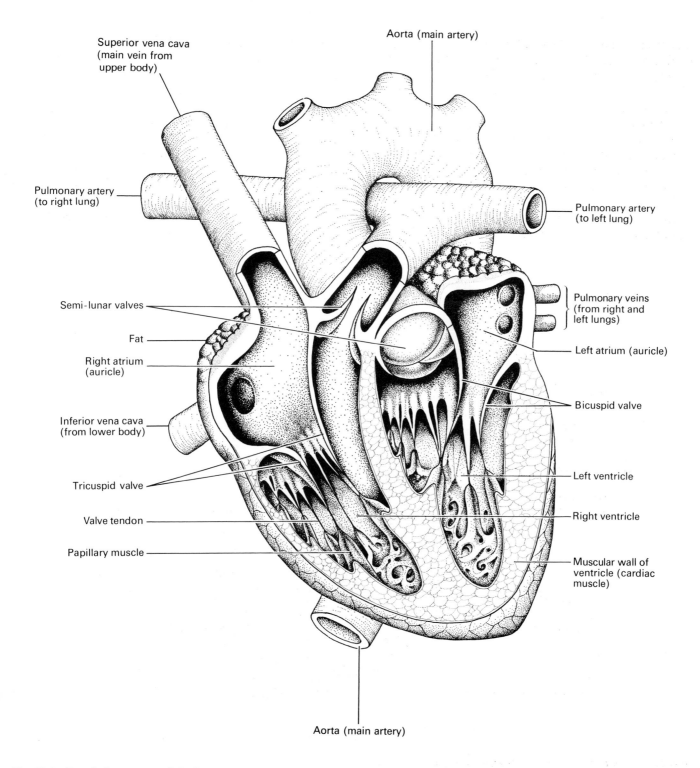

Superior vena cava (main vein from upper body)

Aorta (main artery)

Pulmonary artery (to right lung)

Pulmonary artery (to left lung)

Semi-lunar valves

Fat

Right atrium (auricle)

Pulmonary veins (from right and left lungs)

Left atrium (auricle)

Bicuspid valve

Inferior vena cava (from lower body)

Tricuspid valve

Left ventricle

Right ventricle

Valve tendon

Papillary muscle

Muscular wall of ventricle (cardiac muscle)

Aorta (main artery)

Fig. 33.2   Detailed structure of the heart

Circulation of blood

The heart pumps blood into vessels called **arteries**. Arteries divide into very narrow vessels called **capillaries**. Capillaries eventually join together into **veins**, which carry blood back to the heart.

Arteries

Arteries have thick walls made mainly of muscle and elastic fibres (Fig. 34.2C). Each heart-beat forces blood into arteries at high pressure, which stretches the artery walls outwards. Between heart-beats elastic and muscle fibres in the arteries press inwards on the blood, forcing it away from the heart. Blood must flow in this direction because the semi-lunar valves stop it flowing back into the heart (Fig. 34.1A). This wave of stretching and contraction in arteries causes the 'pulse' which can be felt in the wrist and neck.

Veins

Veins have thinner walls than arteries (Fig. 34.2A). Blood in veins is at low pressure but it flows back to the heart for two reasons. First, the valves in a vein open like double doors, but in only one direction, allowing blood to move *towards* the heart. These valve flaps close if blood starts moving in the opposite direction. Second, the longer veins are situated inside large muscles, for example, in the muscles of the legs and arms. When these muscles contract as the limbs move they squeeze the veins flat. This pressure helps to push blood towards the heart.

Double circulation of blood

Mammals have what is called a double circulatory system (Fig. 34.2), in which the blood flows through two systems. First, blood is pumped from the right ventricle to the lungs, and back to the left atrium of the heart. From here it passes into the left ventricle. Second, blood is pumped from the left ventricle to all other parts of the body, and back to the right atrium of the heart.

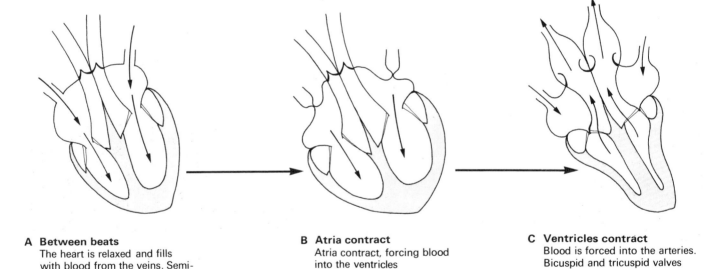

A Between beats
The heart is relaxed and fills with blood from the veins. Semi-lunar valves prevent blood entering through the arteries

B Atria contract
Atria contract, forcing blood into the ventricles

C Ventricles contract
Blood is forced into the arteries. Bicuspid and tricuspid valves prevent blood flowing back into the atria. The atria begin to fill with blood

Fig. 34.1   Diagram showing how the heart pumps blood

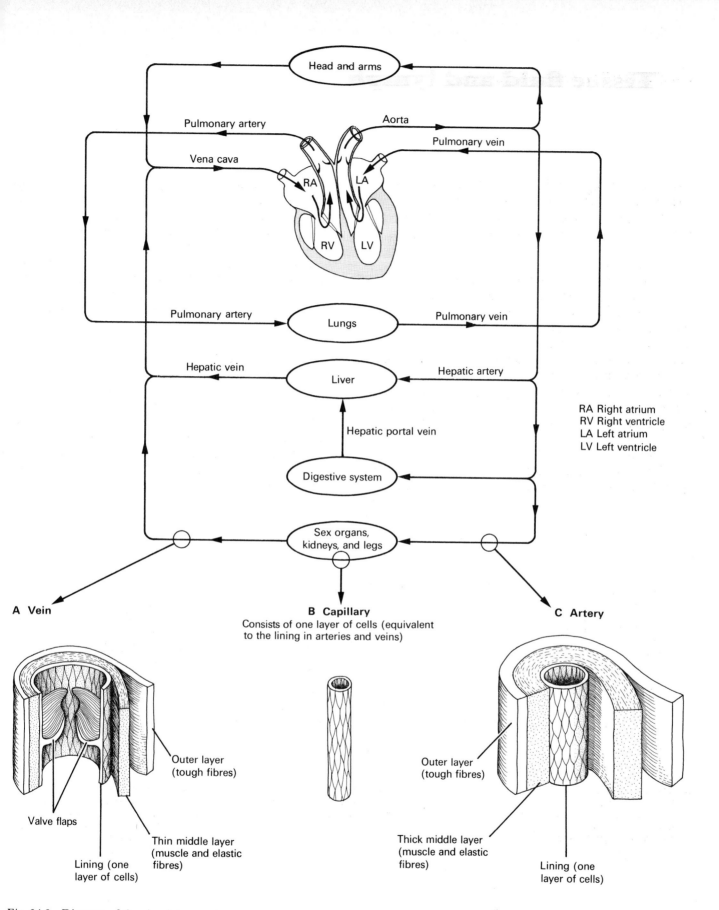

Head and arms

Pulmonary artery

Aorta

Pulmonary vein

Vena cava

RA

LA

RV

LV

Pulmonary artery

Lungs

Pulmonary vein

Hepatic vein

Liver

Hepatic artery

RA Right atrium
RV Right ventricle
LA Left atrium
LV Left ventricle

Hepatic portal vein

Digestive system

Sex organs,
kidneys, and legs

A Vein

B Capillary

Consists of one layer of cells (equivalent
to the lining in arteries and veins)

C Artery

Outer layer
(tough fibres)

Outer layer
(tough fibres)

Valve flaps

Thin middle layer
(muscle and elastic
fibres)

Thick middle layer
(muscle and elastic
fibres)

Lining (one
layer of cells)

Lining (one
layer of cells)

Fig. 34.2   Diagram of the circulatory system

Tissue fluid and lymph

Tissue fluid

Blood containing oxygen and food travels at high pressure through arteries to all parts of the body. Arteries divide into fine networks of capillaries, known as **capillary beds** (Fig. 35.3). Capillary walls are so thin that liquid from the blood passes through them into the spaces between the body cells. This liquid is called **tissue fluid**. As tissue fluid filters through the capillary walls it carries with it oxygen, food, and other useful substances from the blood-stream to the body cells.

As tissue fluid moves between the cells it distributes these useful substances, and at the same time it carries away carbon dioxide and other wastes produced by the cells. After picking up waste products, some tissue fluid passes back into the blood-stream through capillary walls at the vein end of each capillary bed. The remaining tissue fluid drains into lymph vessels, and becomes **lymph**.

Lymph

Lymph vessels begin inside capillary beds, and are about as numerous as blood capillaries (Fig. 35.3). Small lymph vessels drain into larger ones which are similar to veins: they have valves, and are situated between muscles. As the muscles contract the lymph is forced through the vessels.

Lymph vessels flow into **lymph glands** (Fig. 35.3). These contain narrow channels with white blood cells attached to their walls. These white cells destroy bacteria and any other solid particles in the lymph, and so protect the body against infection.

The main lymph vessels join veins near the heart, and the lymph empties into the blood-stream.

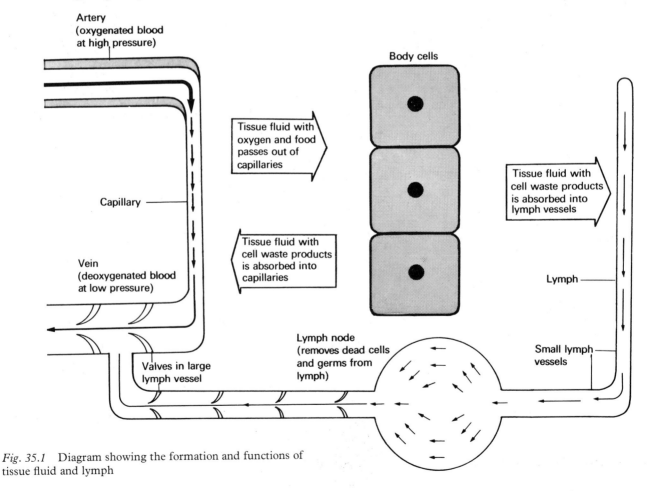

Fig. 35.1 Diagram showing the formation and functions of tissue fluid and lymph

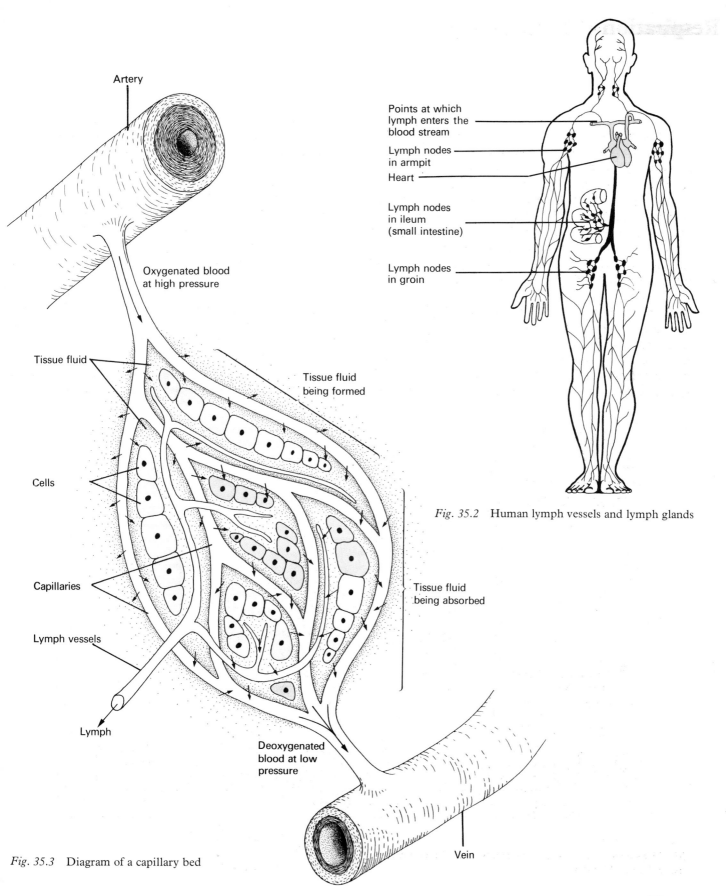

Artery

Oxygenated blood
at high pressure

Tissue fluid

Cells

Capillaries

Lymph vessels

Lymph

Tissue fluid
being formed

Tissue fluid
being absorbed

Deoxygenated
blood at low
pressure

Vein

Fig. 35.3   Diagram of a capillary bed

Points at which
lymph enters the
blood stream

Lymph nodes
in armpit

Heart

Lymph nodes
in ileum
(small intestine)

Lymph nodes
in groin

Fig. 35.2   Human lymph vessels and lymph glands

Respiration

Respiration is the chemical breakdown of food to release the energy which is essential for all living things. Respiration takes place inside the cells and tissues of the body. Consequently, it is often called **cell respiration**, or **tissue respiration**. It is also called **internal respiration** to distinguish it from breathing or **external respiration**, which is simply the movement of air in and out of the lungs.

Most organisms require a constant supply of oxygen to respire. When this oxygen reaches the cells it combines with glucose sugar into which food has been converted. Energy is released, together with waste products of carbon dioxide and water. This chemical reaction can be written as follows:

$$C_6H_{12}O_6 + 6O_2 \longrightarrow 6CO_2 + 6H_2O + Energy$$
Glucose   *Oxygen*   *Carbon dioxide*   *Water*

Respiration which uses oxygen is called **aerobic respiration**. Under certain circumstances energy can be released from food without oxygen. This process is called **anaerobic respiration**. Very little energy is released during anaerobic respiration because glucose is not completely broken down into carbon dioxide and water; it is changed into chemicals such as alcohol and lactic acid. A little carbon dioxide is produced as waste.

Yeast and certain bacteria obtain most of their energy by a type of anaerobic respiration called **fermentation**. Yeast ferments sugar into alcohol, and is used in the production of wines and beer. Fermentation by bacteria produces chemicals such as oxalic, citric, and butyric acids, which have many industrial uses.

Plants can respire anaerobically for several days. This enables them to survive in conditions where animals would suffocate from lack of oxygen. For example, plants can recover after being completely submerged in a flooded field during a period of wet weather.

During very strenuous exercise muscles use up oxygen faster than the blood can supply it. Under these circumstances the muscles change from aerobic to anaerobic respiration; that is, they continue working without using oxygen. This process is explained opposite the photograph below.

During strenuous exercise the heart pumps blood at a rate of about 34 litres per minute. This delivers blood to the muscles at a rate of about 4 litres a minute. But this is not fast enough for the muscles to continue respiring aerobically. They change over to anaerobic respiration, which requires no oxygen, and produces lactic acid and carbon dioxide. Eventually the lactic acid accumulates to a level which prevents muscular contraction, and the muscles stop working. The period of rapid breathing which follows strenuous exercise absorbs the extra oxygen needed to remove lactic acid from the body

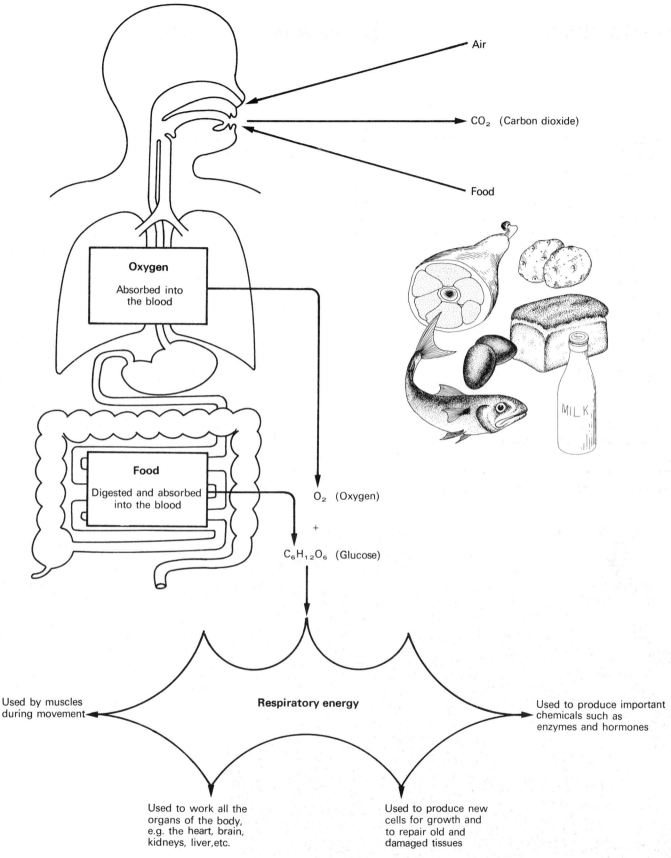

Air

CO_2 (Carbon dioxide)

Food

Oxygen

Absorbed into the blood

Food

Digested and absorbed into the blood

O_2 (Oxygen)

+

$C_6H_{12}O_6$ (Glucose)

Respiratory energy

Used by muscles during movement

Used to produce important chemicals such as enzymes and hormones

Used to work all the organs of the body, e.g. the heart, brain, kidneys, liver, etc.

Used to produce new cells for growth and to repair old and damaged tissues

Fig. 36.1 Diagram of human respiration

The human respiratory system

The respiratory system consists of air passages through the nose, the windpipe or **trachea**, and the lungs. The lungs are situated in a space within the chest called the **thoracic cavity**. The walls of this cavity are supported by the ribs (Fig. 37.1), and its floor is a sheet of muscle called the **diaphragm**.

Air breathed in through the nose is warmed, made moist, and dust and germs are removed before it reaches the lungs. Air passages in the nose and the whole of the windpipe are lined with a carpet of microscopic hairs called **cilia**. Between the cilia are cells which produce a sticky fluid called **mucus** in which germs and dust are trapped. Cilia move back and forth like the oars of a boat, making the mucus with its trapped germs and dust flow to the back of the throat, where it is swallowed.

Air passes through the throat and down the trachea into the lungs. The walls of the trachea are supported by C-shaped rings of cartilage, which ensure that air can flow freely through it, even when the neck is bent. The voice box, or **larynx**, is at the top of the trachea. The larynx contains folds of skin called **vocal cords** which vibrate, causing sounds when air from the lungs is forced over them. At its base, the trachea divides into two tubes called **bronchi**. One bronchus leads into each lung.

Inside the lungs the bronchi divide like the branches of a tree into countless extremely narrow tubes called **bronchioles**. Bronchioles end in tiny bubble-like air-sacs called **alveoli** (Fig. 37.2B). Each alveolus is about 0·2 mm in diameter. Under a microscope the clusters of alveoli at the end of each bronchiole look rather like bunches of grapes. These give the lungs their spongy texture.

A network of blood capillaries surrounds each alveolus (Fig. 37.2C). The capillaries absorb oxygen from the alveoli, and release carbon dioxide into them.

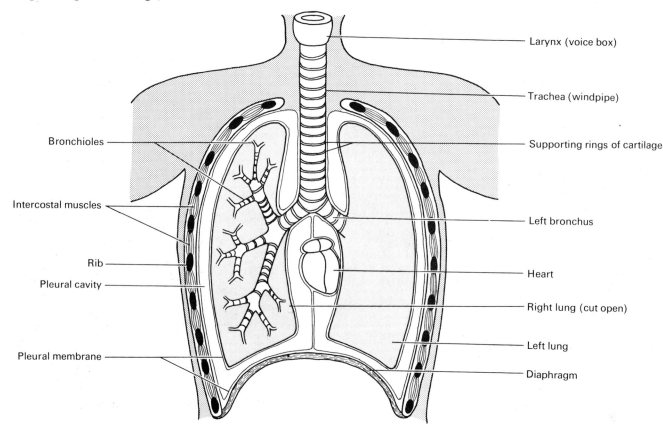

Larynx (voice box)

Trachea (windpipe)

Supporting rings of cartilage

Left bronchus

Heart

Right lung (cut open)

Left lung

Diaphragm

Bronchioles

Intercostal muscles

Rib

Pleural cavity

Pleural membrane

Fig. 37.1   Diagram of the human respiratory system (simplified)

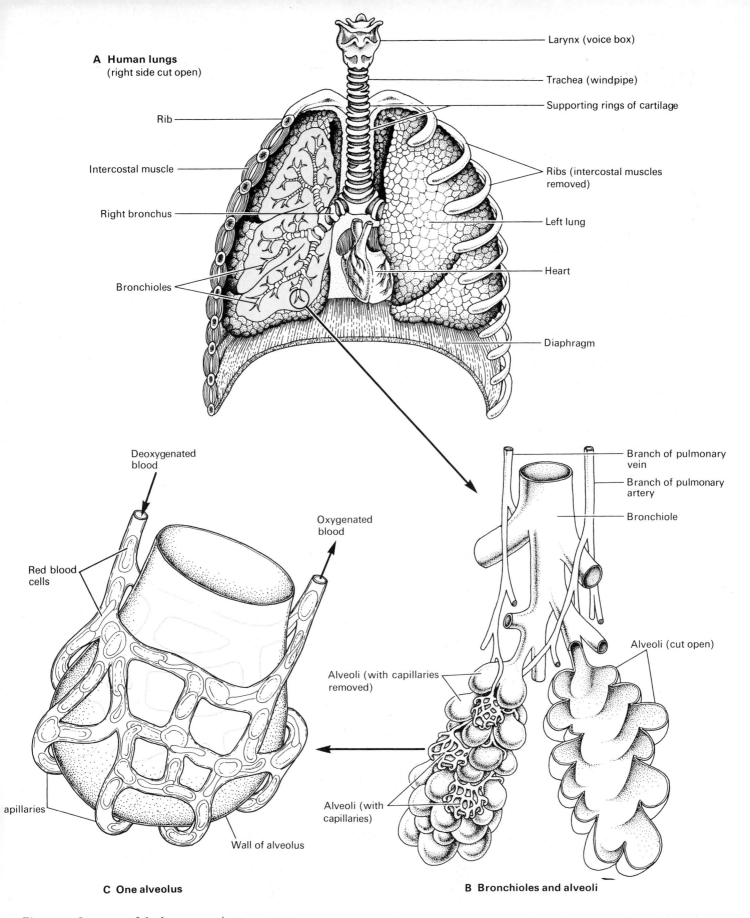

A Human lungs
(right side cut open)

Larynx (voice box)

Trachea (windpipe)

Supporting rings of cartilage

Rib

Intercostal muscle

Ribs (intercostal muscles removed)

Right bronchus

Left lung

Bronchioles

Heart

Diaphragm

Deoxygenated blood

Oxygenated blood

Branch of pulmonary vein

Branch of pulmonary artery

Bronchiole

Red blood cells

Alveoli (with capillaries removed)

Alveoli (cut open)

apillaries

Alveoli (with capillaries)

Wall of alveolus

C One alveolus

B Bronchioles and alveoli

Fig. 37.2   Structure of the human respiratory system

75

Breathing

Air is moved in and out of the lungs by the action of the **diaphragm** and **intercostal muscles**.

Inspiration (breathing in)

The diaphragm is dome-shaped and relaxed before a breath is taken (Fig. 38.1B). During the first stage of breathing in the diaphragm contracts and becomes flatter in shape (Fig. 38.1A). At the same time **external intercostal muscles** (Fig. 38.4) contract, pulling the ribs upwards so that they pivot where they join the backbone (Fig. 38.3). The flattening of the diaphragm and the lifting of the rib cage increase the volume of the thoracic cavity. This is automatically followed by an increase in lung volume. The increase in volume reduces air pressure inside the lungs, and air rushes into them from the atmosphere through the air passages.

Expiration (breathing out)

The diaphragm and external intercostal muscles relax. The rib cage drops and the diaphragm returns to its original dome shape. These movements reduce the volume of the thoracic cavity and the lungs return to their original volume. This squeezes air out of the lungs through the air passages. Air can be forced out of the lungs by contracting the **internal intercostal muscles**. This happens, for example, when playing a wind instrument.

At rest, an adult breathes about 16–18 times a minute, using the diaphragm alone. During exercise this rate increases, and the intercostal muscles are used to increase the volume of each breath. This supplies the muscles with extra oxygen and removes carbon dioxide more quickly.

Figure 38.2 demonstrates how the intercostal muscles work. Fit an elastic band through holes A and B and watch how the length changes as you move the apparatus. The elastic band represents the external intercostal muscles (which raise the rib cage). Then fit the elastic band through holes C and D. It now represents the internal intercostal muscles (which lower the rib cage, forcing air from the lungs).

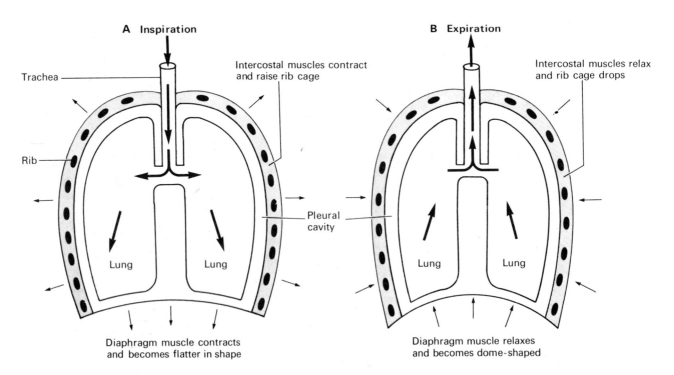

Fig. 38.1 Front view of thorax, showing rib and diaphragm movements

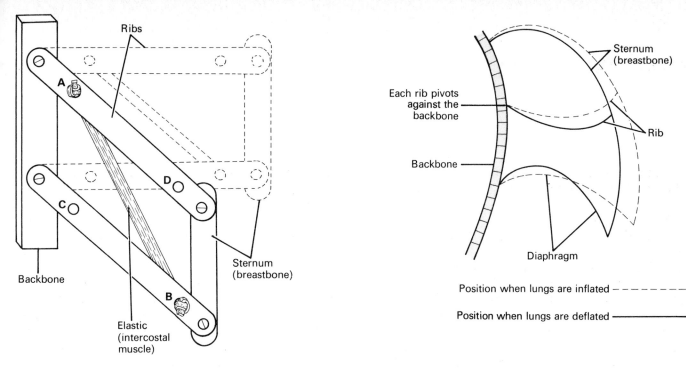

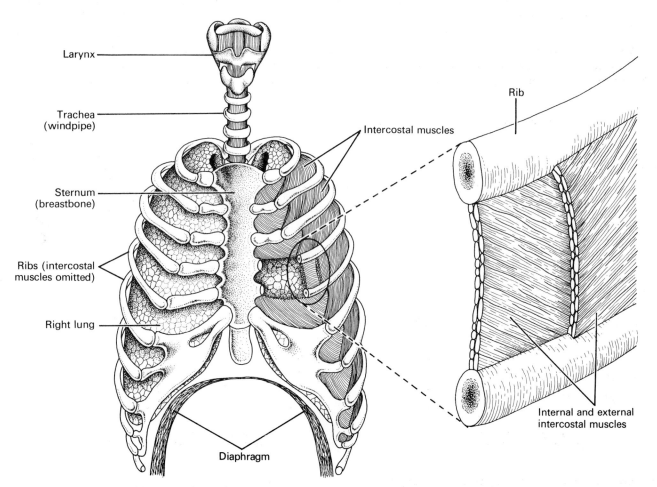

Fig. 38.2 Model showing how intercostal muscles move the rib cage

Fig. 38.3 Side view of the rib cage showing movements during breathing

Fig. 38.4 Rib cage and intercostal muscles

Gaseous exchange

Most organisms absorb oxygen into their bodies, and at the same time release carbon dioxide from their bodies. In other words they 'exchange gases' with their surroundings. This process is called **gaseous exchange**.

In most animals gaseous exchange takes place at a specialized region of the body called a **respiratory surface**. In fish, for example, the respiratory surface is the gills, and in humans it is the alveoli of the lungs (Fig. 39.2).

Absorption of oxygen

Blood entering the capillaries of the lungs is **deoxygenated** (that is, it contains no oxygen), because haemoglobin in its red cells has released all its oxygen to the cells of the body. But air breathed from the atmosphere into the lung alveoli is rich in oxygen. This oxygen dissolves in a film of water lining the alveoli (Fig. 39.2B), then diffuses through the alveoli and capillary walls (a distance of only 0·005 mm) into the blood.

The capillaries that cover the alveoli walls are narrower in diameter than red blood cells. The red cells are squeezed and change shape as they are forced through the lung capillaries by blood pressure. This helps oxygen absorption in two ways. First, a greater area of each red cell is pressed against capillary walls through which oxygen is diffusing. Second, red cells are slowed down as they squeeze through the capillaries. This increases the time available for oxygen absorption. When blood leaves the lung capillaries it is **oxygenated**.

Release of carbon dioxide

Blood entering lung capillaries is full of carbon dioxide, dissolved mainly in the plasma. This gas has been absorbed by the blood from body cells. Carbon dioxide diffuses out of the blood through the capillary and alveoli walls into the film of water, and then into the alveoli. Finally, it is removed from the lungs during expiration (breathing out).

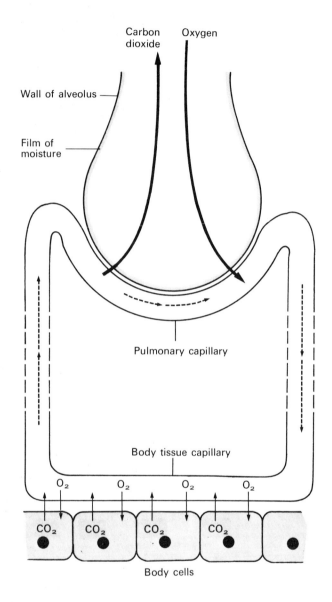

Fig. 39.1 Diagram of gaseous exchange between an alveolus and the blood-stream, and between blood and body cells

A An alveolus cut open

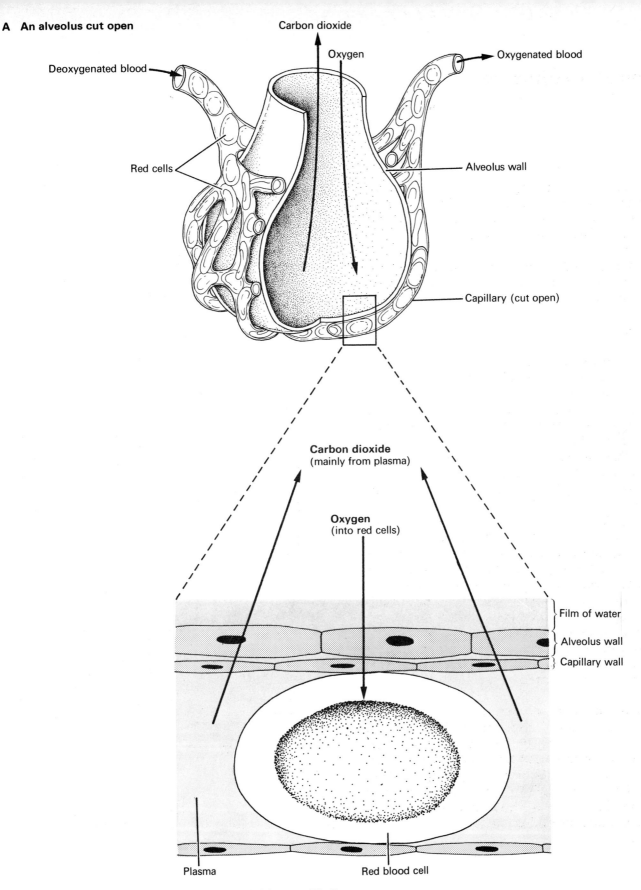

Carbon dioxide

Oxygen

Deoxygenated blood

Oxygenated blood

Red cells

Alveolus wall

Capillary (cut open)

Carbon dioxide
(mainly from plasma)

Oxygen
(into red cells)

Film of water

Alveolus wall

Capillary wall

Plasma

Red blood cell

B Cross-section of alveolus and capillary (highly magnified)

Fig. 39.2 Gaseous exchange in an alveolus

Respiratory organs in fish

Fish absorb oxygen from water, and release carbon dioxide into water, by means of their gills. In bony fish there are four gills on each side of the body, and each set is situated behind a flap of skin called an **operculum** (*plural*: **opercula**). A gill consists of a bone called the **gill bar**, and hundreds of tiny flaps of skin called **gill lamellae** (Fig. 40.2A and B). The lamellae are supported by the bone, and are situated one on top of the other. This allows water to flow between them during breathing movements. The surface of each lamella has projecting ridges covered with a network of capillaries (Fig. 40.2C). Gaseous exchange takes place between the blood in these capillaries and the water which flows over them.

Breathing movements

A fish uses its mouth, muscles in the floor of its mouth cavity, and its opercula to produce a continuous flow of water over its gills. Figure 40.1 and the following notes explain how this is done:

1. The mouth opens and at the same time muscles lower the floor of the mouth cavity. This reduces water pressure inside the mouth, and water flows into it.

2. Muscles then make the opercula bulge outwards. This reduces pressure in the gill chamber, causing water to flow from the open mouth to the gills. During this stage the edge of each operculum is pressed tightly against the sides of the body by the higher pressure of the water outside, so water enters the fish only through its mouth.

3. The mouth then closes, and muscles raise the floor of the mouth cavity. This squeezes the remaining water in the mouth towards the gills.

4. Muscles squeeze the walls of the opercula inwards so that pressure around the gills becomes higher than pressure outside the body. This lifts the edge of each operculum, letting water flow between the gills and out of the body.

The process is then repeated.

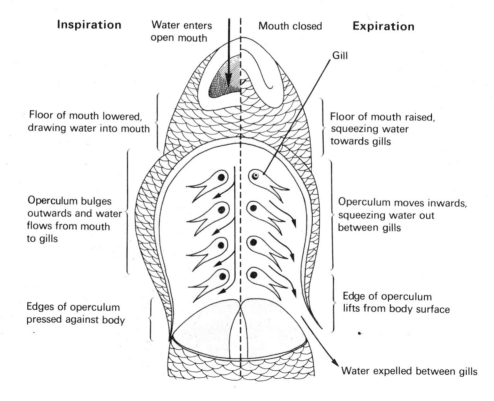

Fig. 40.1 Diagram showing breathing movements in fish

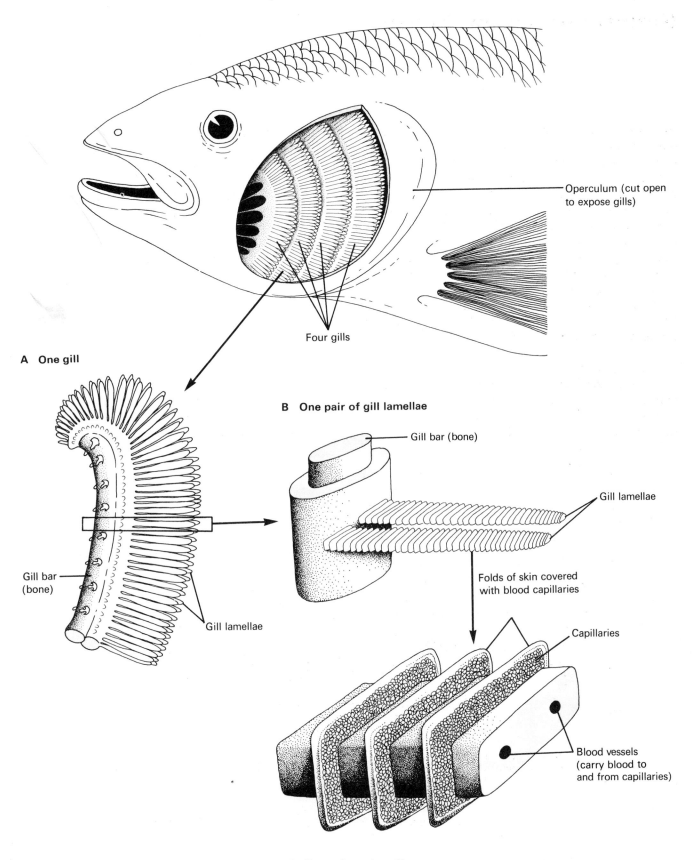

Operculum (cut open
to expose gills)

Four gills

A One gill

B One pair of gill lamellae

Gill bar (bone)

Gill lamellae

Gill bar
(bone)

Gill lamellae

Folds of skin covered
with blood capillaries

Capillaries

Blood vessels
(carry blood to
and from capillaries)

Fig. 40.2   Structure of a gill

C Part of one lamella

41

Respiratory organs in insects

In humans and in fish, oxygen and carbon dioxide are transported between body cells and respiratory organs by the blood. But in insects these gases are transported directly to and from the cells through a network of fine tubes which open to the air at the body surface (Fig. 41.1). This network of tubes is called the **tracheal system**, and the pores through which they open to the air are called **spiracles**.

There is usually a pair of spiracles on each segment of the body. Spiracles open into tracheal tubes, the walls of which are strengthened by spiral folds of cuticle (Fig. 41.2). These folds prevent the tubes from being flattened, and allow them to be extended and contracted rather like an accordion during breathing.

Tracheal tubes from each spiracle join other tubes running along the insect on each side of its body. These join still more tubes which pass deep into the body, dividing like roots until they form extremely narrow **tracheoles** only 0·002 mm in diameter. Gaseous exchange takes place through the walls of the tracheoles.

Air is drawn into and squeezed out of the tracheal system by movements of the insect's abdomen. In bees and wasps segments of the abdomen slide in and out like a telescope. But in cockroaches, locusts, and beetles muscles squeeze the abdomen flat and then relax, thus letting it fill out again.

In some insects the tracheal system contains thin-walled air-sacs. These are squeezed and expanded during breathing movements, thus increasing the volume of air moving in and out of the tracheal system. Some air-sacs are situated between layers of flight muscle, and are compressed and expanded as the insect flaps its wings. This automatically increases the rate of gaseous exchange during flight.

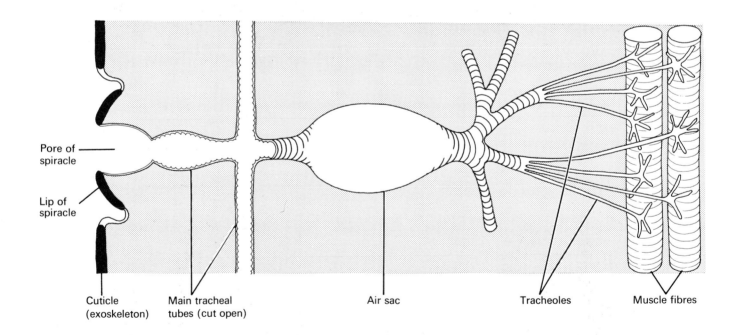

Pore of spiracle

Lip of spiracle

Cuticle (exoskeleton)

Main tracheal tubes (cut open)

Air sac

Tracheoles

Muscle fibres

Fig. 41.1 Diagram of a cross-section through the main parts of a insect's respiratory system

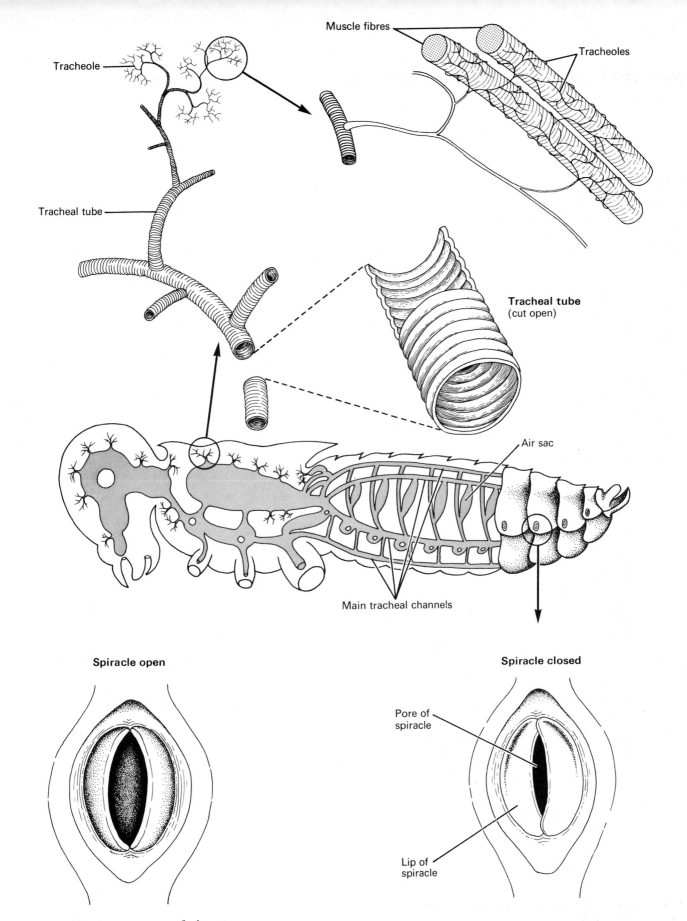

Tracheole

Muscle fibres

Tracheoles

Tracheal tube

Tracheal tube
(cut open)

Air sac

Main tracheal channels

Spiracle open

Spiracle closed

Pore of
spiracle

Lip of
spiracle

Fig. 41.2 Respiratory system of a locust

The skin and temperature control

Homoiothermic animals can maintain a fairly constant body temperature despite temperature changes in their surroundings. Only birds and mammals are homoiothermic. All other animals are **poikilothermic**, which means their body temperature varies according to the temperature of their surroundings.

Prevention of over-heating

Homoiothermic animals, e.g. humans, are in danger of over-heating in hot weather, and during exercise and illness when their bodies produce excess heat. Over-heating is prevented in the following ways (Fig. 42.1B).

1. *Sweating*   Sweat is a watery liquid produced by sweat glands (Fig. 42.2). As sweat evaporates from the skin it removes heat from the body.

2. *Vasodilation*   Vasodilation is the enlargement of blood vessels so that more blood flows through them. If the body becomes over-heated vasodilation occurs in the **superficial capillaries** of the skin (Fig. 42.1B). The over-heated blood then flows close to the body surface where it rapidly loses heat.

Prevention of over-cooling

1. *Shivering*   In cold weather the muscles begin jerky rhythmic movements called shivering. Muscles respire at a faster rate during shivering and generate more heat to keep the body warm.

2. *Vasoconstriction*   This is the opposite of vasodilation: blood vessels become narrower so that less blood flows through them. In cold weather vasoconstriction of the superficial blood vessels reduces loss of heat by radiation through the skin (Fig. 42.1A).

3. *Contraction of hair erector muscles*   In cold weather hair erector muscles contract, making the hairs 'stand on end' (Fig. 42.1A). The upright hairs prevent cold air from reaching the skin. Still air trapped between the upright hairs is heated by the body and forms a thick warm 'blanket' around the whole animal.

4. *Fat*   In cold seasons many animals develop a thick layer of fat beneath the skin (Fig. 42.1A). This fat insulates the body, which means it retains more heat. Fat is also a store of food.

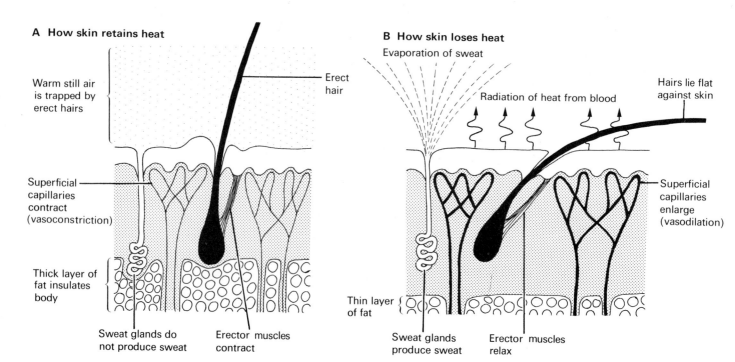

Fig. 42.1   Diagram showing how skin controls temperature

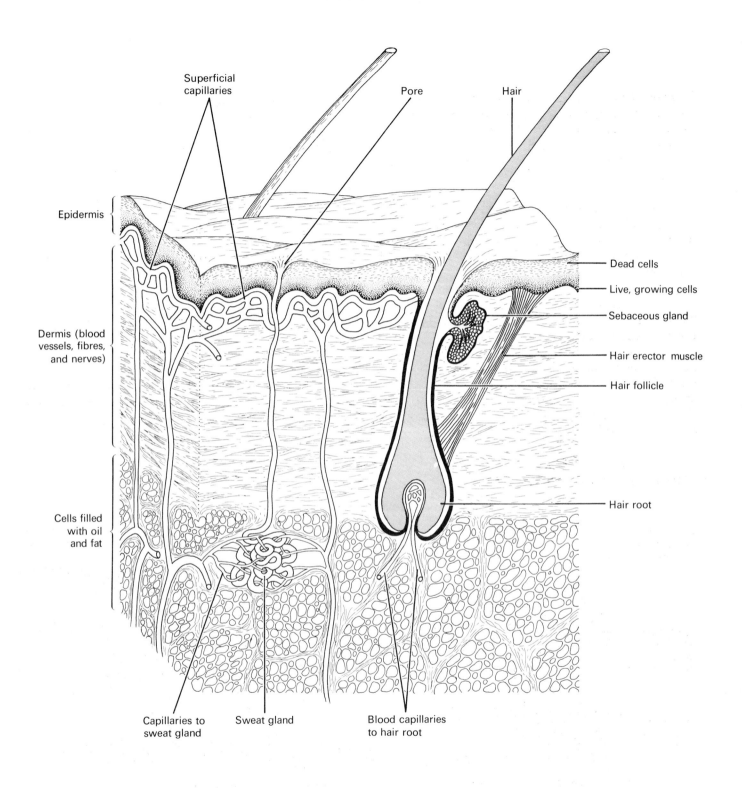

Superficial capillaries

Pore

Hair

Epidermis

Dead cells

Live, growing cells

Sebaceous gland

Dermis (blood vessels, fibres, and nerves)

Hair erector muscle

Hair follicle

Cells filled with oil and fat

Hair root

Capillaries to sweat gland

Sweat gland

Blood capillaries to hair root

Fig. 42.2   Parts of the skin concerned with temperature control

Excretion by the kidneys

Excretion is the removal from the body of waste and unwanted substances produced by the processes of life. Carbon dioxide, for example, is produced by respiration, and is then excreted by the lungs. **Urea** is a waste substance produced by the liver. As the body assimilates proteins it produces poisonous substances. The liver converts these into urea, which is less poisonous. Urea is then excreted by the kidneys in a liquid called **urine**.

Kidneys are part of a set of organs called the **urinary system** (Fig. 43.1). Each kidney receives blood through a **renal artery**. Inside a kidney this artery divides into capillaries which carry blood to over a million tiny structures called **Bowman's capsules** (Fig. 43.3B). Inside a Bowman's capsule the capillary divides again, forming a ball of intertwined capillaries called a **glomerulus** (*plural:* glomeruli).

Blood pressure in a glomerulus is so high that liquid is forced out through its walls into the space within the Bowman's capsule. This liquid is called **glomerular filtrate**, because it is formed by blood filtering through the glomerulus wall, and the inner wall of the Bowman's capsule. Glomerular filtrate contains urea, but it also contains many useful substances such as glucose, amino acids, vitamins, and minerals.

Glomerular filtrate flows out of the Bowman's capsules into long, coiled **kidney tubules** (Fig. 43.3A). Here **reabsorption** takes place: all the *useful* substances (glucose, etc.) are reabsorbed into the blood-stream. Reabsorption leaves behind a liquid called urine, which consists of urea and unwanted minerals dissolved in water.

Urine flows down to the bladder through the **ureters** (Fig. 43.1). The bladder expands as it fills until it is stretched to the point at which nerve endings in its walls send impulses to the brain. These impulses act as a signal that the bladder is full. A sphincter muscle at the base of the bladder is voluntarily relaxed to let urine flow out of the body. This is called **urination**.

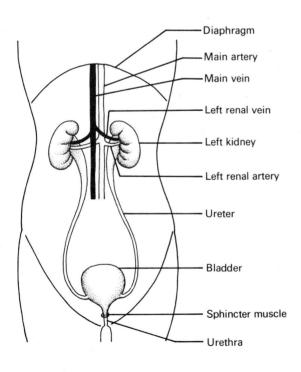

Diaphragm

Main artery

Main vein

Left renal vein

Left kidney

Left renal artery

Ureter

Bladder

Sphincter muscle

Urethra

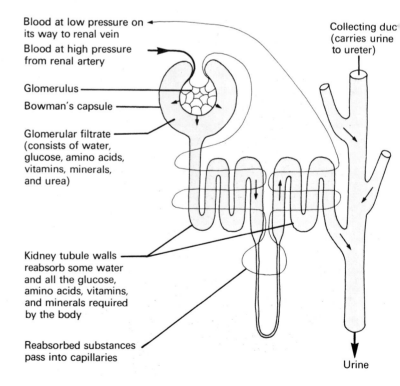

Blood at low pressure on its way to renal vein

Blood at high pressure from renal artery

Collecting duc (carries urine to ureter)

Glomerulus

Bowman's capsule

Glomerular filtrate (consists of water, glucose, amino acids, vitamins, minerals, and urea)

Kidney tubule walls reabsorb some water and all the glucose, amino acids, vitamins, and minerals required by the body

Reabsorbed substances pass into capillaries

Urine

Fig. 43.1   The human urinary system

Fig. 43.2   Diagram showing how urine is formed

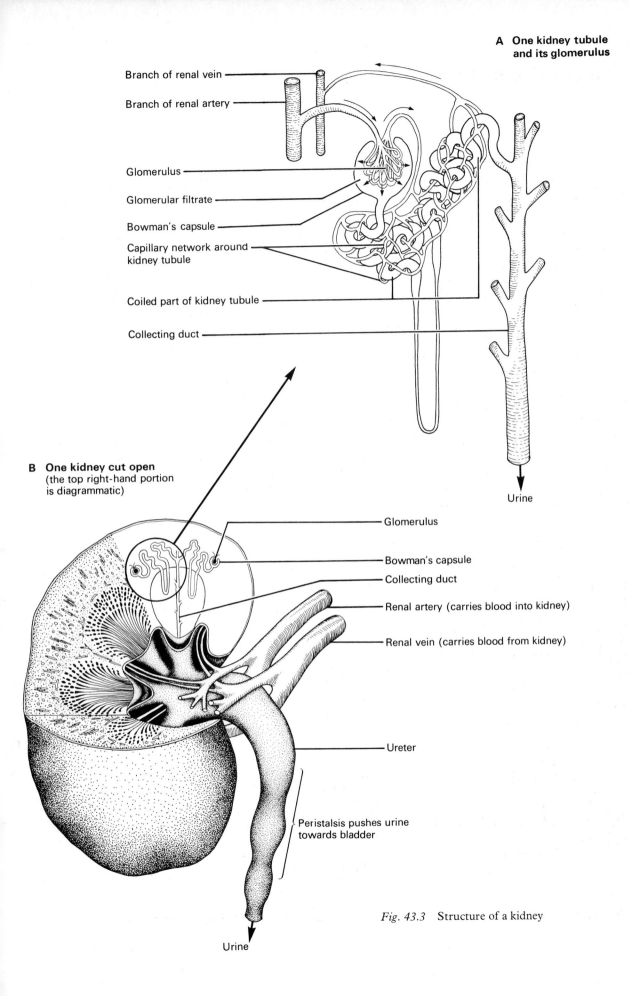

A One kidney tubule and its glomerulus

Branch of renal vein

Branch of renal artery

Glomerulus

Glomerular filtrate

Bowman's capsule

Capillary network around kidney tubule

Coiled part of kidney tubule

Collecting duct

Urine

B One kidney cut open
(the top right-hand portion is diagrammatic)

Glomerulus

Bowman's capsule

Collecting duct

Renal artery (carries blood into kidney)

Renal vein (carries blood from kidney)

Ureter

Peristalsis pushes urine towards bladder

Urine

Fig. 43.3 Structure of a kidney

Senses of the skin, nose, and tongue

Skin

Skin has nerve endings sensitive to touch, pressure, pain, and temperature (Fig. 44.1). These, and other sensory nerve endings, are called **receptors** because they 'receive' stimulation from the outside world.

Touch and pressure In humans touch and pressure receptors are concentrated in the skin of the tongue and fingertips. These receptors detect the texture of objects: whether they are rough, smooth, hard, or soft, for instance. Touch receptors are attached to hairs. If an object brushes against any of the hairs which emerge from the skin, the receptors are stimulated.

Pain Pain receptors are more evenly distributed over the skin, and are also found in most tissues and organs inside the body. Pain acts as a warning signal. It tells the brain that something is wrong with the body.

Temperature There are separate 'heat' and 'cold' receptors in the skin. These are used to detect changes in temperature. The fingertips can detect temperature differences as small as 0·5 °C.

Nose

Smell receptors, or **olfactory organs**, are sensitive to chemicals in the air (Fig. 44.3A). But these chemicals must first dissolve in the film of moisture which covers the receptors.

Tongue

The taste receptors, or **taste buds**, in the tongue are also sensitive to chemicals (Fig. 44.3B and C). There are four types of taste bud: those sensitive to salt, sweet, sour, and bitter-tasting substances. The countless different flavours of food and drink are identified according to how much they stimulate these four types of receptor. Groups of each type of receptor are concentrated in certain areas of the tongue (Fig. 44.2).

Certain tastes and smells are extremely unpleasant. These sensations are usually produced by poisonous or decaying substances, which could be harmful. These receptors therefore protect the body from harm in the same way as pain receptors.

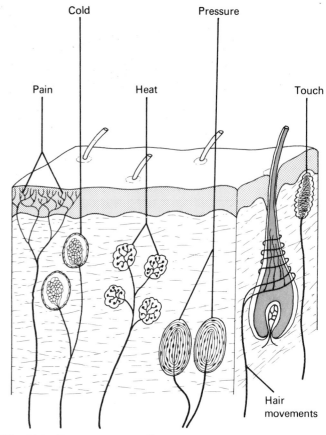

Fig. 44.1   Sensory nerve endings in the skin

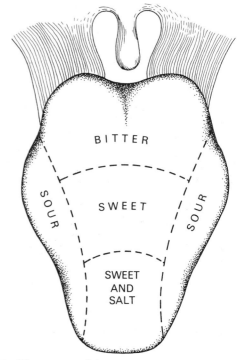

Fig. 44.2   Taste areas of the tongue

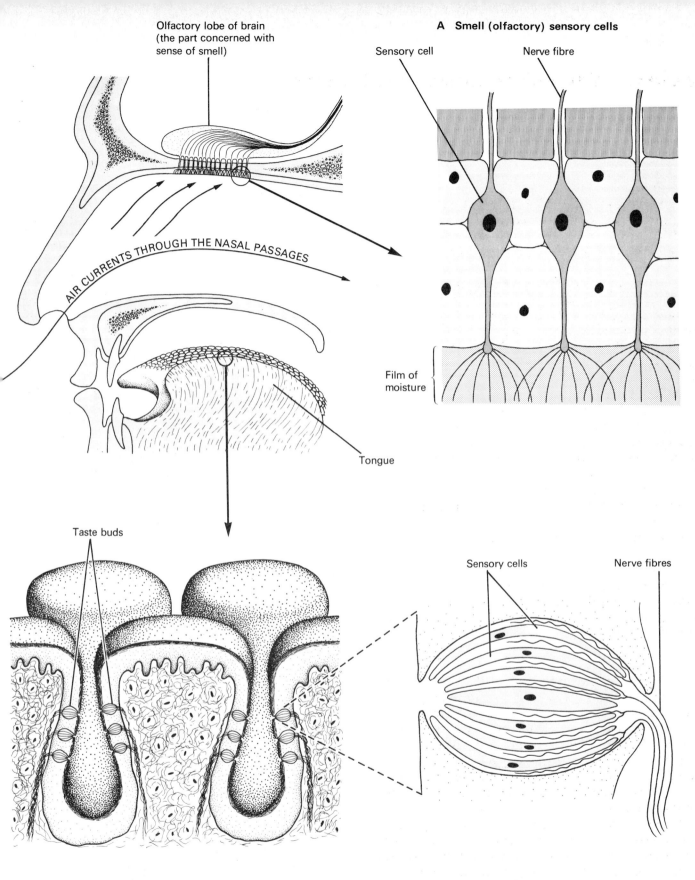

Olfactory lobe of brain
(the part concerned with
sense of smell)

A Smell (olfactory) sensory cells

Sensory cell

Nerve fibre

AIR CURRENTS THROUGH THE NASAL PASSAGES

Tongue

Film of
moisture

Taste buds

Sensory cells

Nerve fibres

B Part of the tongue magnified to show taste buds

C One taste bud

Fig. 44.3   Senses of the nose and tongue

179423

Eyes and vision

An eye is a hollow ball made of tough white fibres called the **sclerotic layer** (Fig. 45.2). At the front of the eyeball is a circular window called the **cornea**. This is covered by a transparent skin called the **conjunctiva**. A black substance called the **choroid** lines the inside of the eyeball. The choroid prevents light from being reflected inside the eyeball. It also contains blood vessels.

The amount of light entering the eye is controlled by the **iris**. The iris is a disc of muscles with a hole at its centre called the **pupil**. In bright light one set of muscles in the iris contracts, making the pupil smaller, and in dim light another set of iris muscles contracts, making it larger.

The cornea and lens of the eye focus a clear, upside-down picture on the back of the eye. The picture, or **image**, falls on a layer of light-sensitive nerve endings called the **retina**. Nerve endings in the retina send impulses to the brain. The brain interprets the impulses from the retina, and the upside-down image is turned the right way round.

The cornea alone can form an image of distant objects on the retina. A clear, oval structure called the **lens** is necessary to focus near objects. The lens is made of an elastic substance and is held in place by **suspensory ligaments**. The shape of the lens is changed depending how far away the object that is being observed is. **Ciliary muscles** alter the shape of the lens, and so control its focusing power.

To focus on a near object, the ciliary muscles contract, forming a circle with a smaller diameter (Fig. 45.3B). This reduces tension on the suspensory ligaments, and so the elastic substance of the lens bulges outwards, forming a fatter or more convex shape. This change of shape increases the power of the lens to bend light, so that near objects are brought into focus. The ciliary muscles contract and squeeze inwards on the **aqueous** and **vitreous humours** (the transparent substances which fill the eyeball). This increases the pressure inside the eye.

To focus on a distant object the ciliary muscles relax. Pressure in the eyeball automatically stretches these muscles outwards into a larger circle. This pulls on the suspensory ligaments, stretching the lens into a slim (less convex) shape. Light is bent less and distant objects are brought into focus (Fig. 45.3A).

A Normal sight:
both distant and near objects can be focused on the fovea

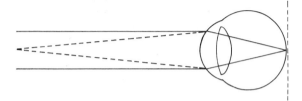

B Long sight (hypermetropia):
occurs when the eyeball is shorter than normal. Distant objects can be focused properly, but the point of focus for an object close to the eye is behind the retina

Long sight is corrected by a converging lens

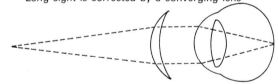

C Short sight (myopia):
occurs when the eyeball is longer than normal. Objects close to the eye can be focused properly, but the point of focus for distant objects is in front of the retina

Short sight is corrected by a diverging lens

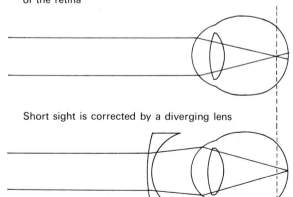

Fig. 45.1 Some eye defects and their correction

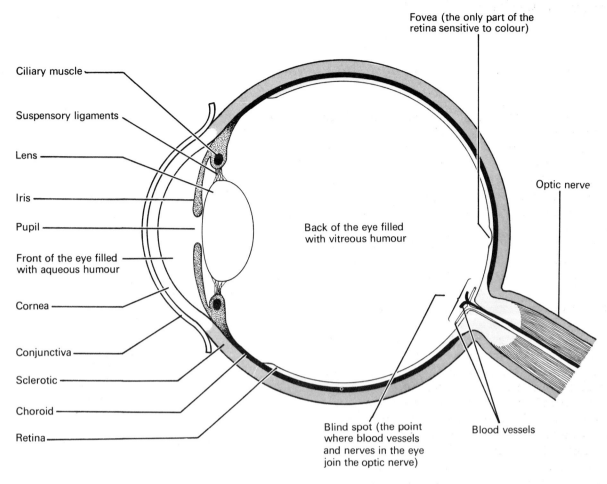

Fovea (the only part of the retina sensitive to colour)

Ciliary muscle

Suspensory ligaments

Lens

Iris

Pupil

Front of the eye filled with aqueous humour

Cornea

Conjunctiva

Sclerotic

Choroid

Retina

Back of the eye filled with vitreous humour

Optic nerve

Blind spot (the point where blood vessels and nerves in the eye join the optic nerve)

Blood vessels

Fig. 45.2 Structure of the eye

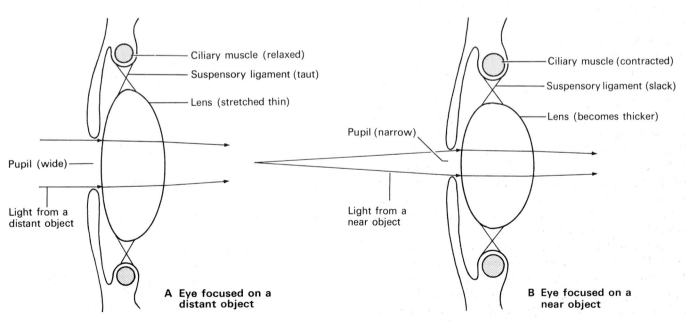

Ciliary muscle (relaxed)

Suspensory ligament (taut)

Lens (stretched thin)

Pupil (wide)

Light from a distant object

A Eye focused on a distant object

Ciliary muscle (contracted)

Suspensory ligament (slack)

Lens (becomes thicker)

Pupil (narrow)

Light from a near object

B Eye focused on a near object

Fig. 45.3 Diagram showing changes in the eye during focusing (accommodation)

Ears and hearing

Sound waves travelling through the air are collected by the funnel-shaped **pinna** of each ear (Fig. 46.2). The pinna directs sound waves down a short tube, the end of which is closed off by a sheet of skin and muscle called the **ear drum**. Sound waves make the ear drum vibrate in and out.

The structures described so far are part of the **outer ear**. Behind the ear drum is an air-filled space called the **middle ear**. Eustachian tubes connect this space with the back of the mouth. These tubes open during swallowing, letting air in and out of the middle ear so that air pressure is always the same on both sides of the ear drum.

A chain of three tiny bones called the **ear ossicles** connect the ear drum with another sheet of skin called the **oval window**. This covers a tiny hole in the bones of the skull opposite the ear drum. When the ear drum vibrates, the ear ossicles move against each other in such a way that they lever the oval window in and out. These movements cause vibrations to pass into the **inner ear**.

The inner ear is a complicated arrangement of tubular passages in the skull bones filled with a liquid called **perilymph**. Inside these passages are smaller tubes filled with **endolymph**. The tube concerned with hearing is the **cochlea**. It is coiled like the shell of a snail.

When the oval window moves in and out it sends vibrations through the perilymph. These vibrations cause part of the cochlea to shake up and down (Fig. 46.1). These shaking movements stimulate sensory nerve endings attached to the cochlea, causing them to send nerve impulses to the brain

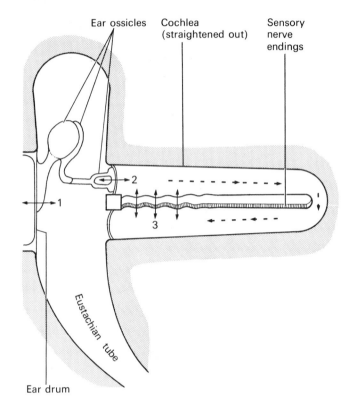

Fig. 46.1 Diagram showing how the ear works (the cochlea has been straightened out)

along the **auditory nerve**. By analysing these impulses the brain discovers many things about sounds: for example, how loud they are, what pitch they are, what direction they come from, and what is causing them.

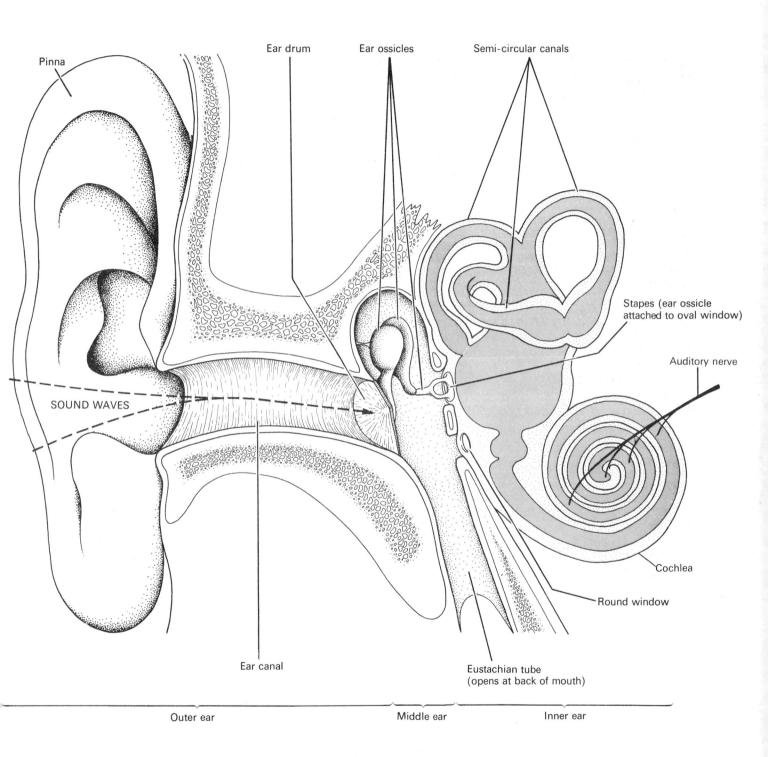

Pinna

Ear drum

Ear ossicles

Semi-circular canals

Stapes (ear ossicle attached to oval window)

Auditory nerve

SOUND WAVES

Cochlea

Round window

Ear canal

Eustachian tube (opens at back of mouth)

Outer ear

Middle ear

Inner ear

Fig. 46.2 Structure of the ear (the middle and inner ears are drawn to a larger scale than the outer ear)

47

Sense of balance

The **semi-circular canals, utricles,** and **saccules** of the inner ear are concerned with balance.

There are three semi-circular canals in each ear. These are curved tubes filled with endolymph, and their function is to detect changes in the direction of movement. Two of them are upright (vertical) and one is horizontal (Fig. 47.1).

At one end of each semi-circular canal is a swelling called an **ampulla,** which contains sensory nerve cells attached to sensory nerve endings. The hairs of these cells are embedded in a cone of jelly called a **cupula** (Fig. 47.2). When the head moves it causes endolymph to flow through one or more of the semi-circular canals. The endolymph presses against a cupula, and bends it over. This stretches the sensory hair cells and stimulates the nerve endings to send impulses to the brain. The brain calculates the direction of the movement from the amount of stimulation it receives from each ampulla.

The utricles and saccules are spaces in the skull filled with endolymph. They detect changes in the speed of movement (acceleration and deceleration), and changes in the position of the body (whether it is upright or tilted).

The inner surface of each utricle and saccule consists of patches of sensory hair cells and nerve endings. The hairs of these cells are embedded in a jellyish substance containing tiny pieces of chalk called **otoliths.** When the head is upright the otoliths press down on the sensory nerves, but when the head tilts, or moves sideways in any direction, the weight of the otoliths pulls against the nerve endings. These then send impulses to the brain. The brain calculates the angle of tilt, or the speed of the movement, from the amount of stimulation from each utricle and saccule.

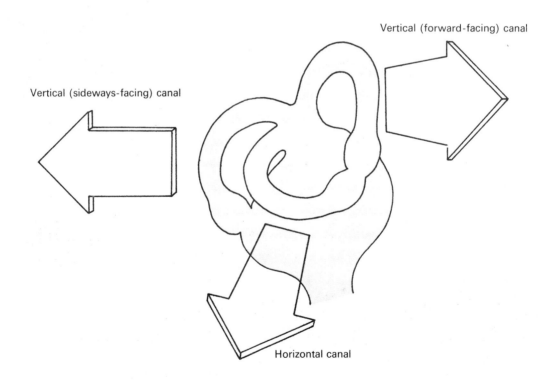

Fig. 47.1   Diagram showing the positions of the semi-circular canals

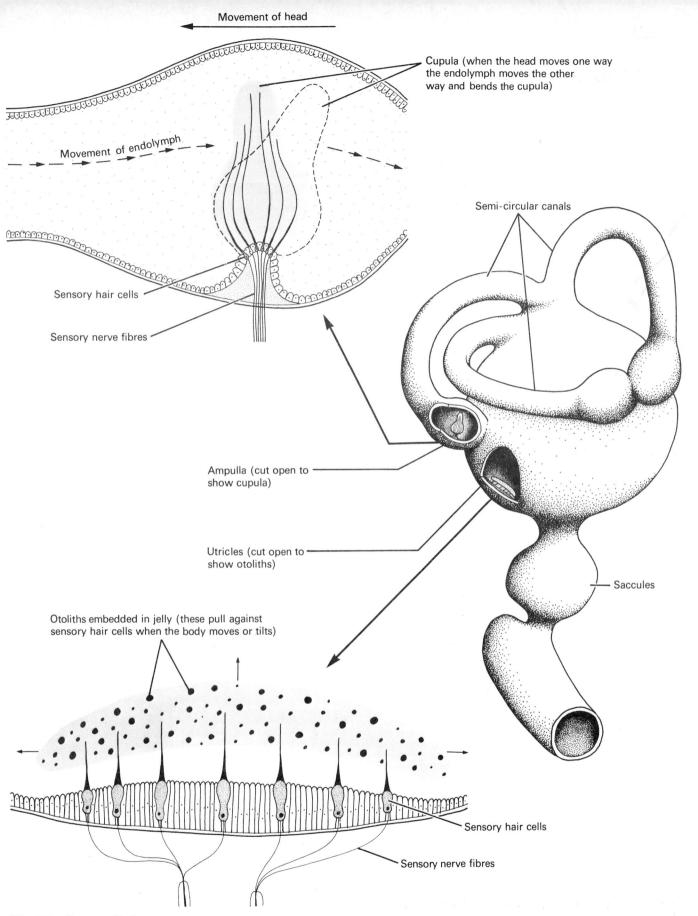

Movement of head

Cupula (when the head moves one way
the endolymph moves the other
way and bends the cupula)

Movement of endolymph

Sensory hair cells

Sensory nerve fibres

Semi-circular canals

Ampulla (cut open to
show cupula)

Utricles (cut open to
show otoliths)

Saccules

Otoliths embedded in jelly (these pull against
sensory hair cells when the body moves or tilts)

Sensory hair cells

Sensory nerve fibres

Fig. 47.2 Organs of balance

48

Phototropism

Plants cannot move from place to place. They can only 'move' by growing in certain directions. Growth movements are called **tropisms** or **tropic responses**. Plant stems, for example, demonstrate **phototropism** when they grow towards light.

The direction of a phototropic movement depends on the direction from which light is coming. Figure 48.1A shows how a bean plant grown out of doors grows straight upwards towards light coming from above. If a plant is grown in the dark it develops a long, thin stem with small yellow leaves (Fig. 48.1B). But if a plant is grown in light coming from one side its stem curves towards the light, and its leaves spread at right angles to the light rays (Fig. 48.1C). This type of growth enables a seedling growing in a shady place to grow a long stem which may reach the light before the food stored in the seed has been used up.

Figure 48.2A shows how to demonstrate phototropism by putting a plant in a cardboard box which has a hole cut in one side. Figure 48.2B is a control experiment. A plant is placed on a **clinostat**, a device which slowly rotates so that the plant is illuminated equally from all sides. In the control experiment the plant grows upwards.

The experiment illustrated in Figure 48.3 shows that only the tip of a plant shoot is sensitive to light. The tip of grass seedling A is covered with black paper. The shoot grows straight upwards even when illuminated from one side (drawing A_1). Seedling B has all of its shoot *except* the tip covered with black paper, and seedling C is completely uncovered. Both these seedlings bend over and grow towards the light (drawings B_1 and C_1).

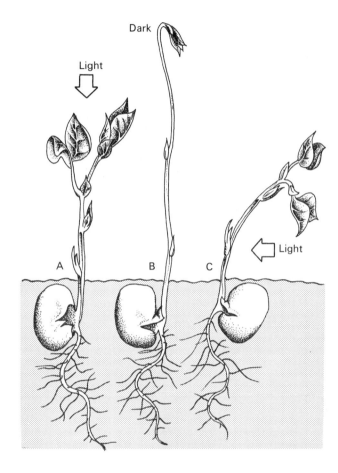

Fig. 48.1 The effect of light on the growth of broad bean plants: **A** was grown out of doors; **B** was grown in the dark; **C** was illuminated from one side

A Plant illuminated from one side
(plant curves towards light)

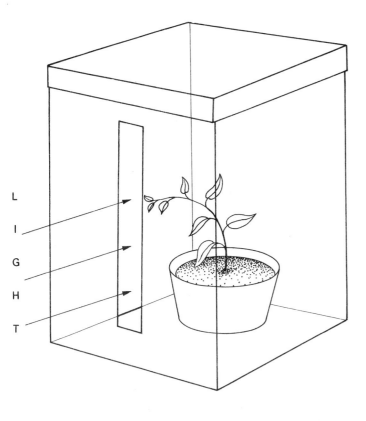

B Plant illuminated from one side while rotating on a *clinostat* (plant grows upwards)

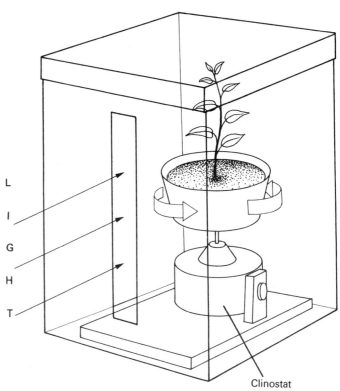

L
I
G
H
T

L
I
G
H
T

Clinostat

Fig. 48.2   How to demonstrate phototropism

Start of experiment

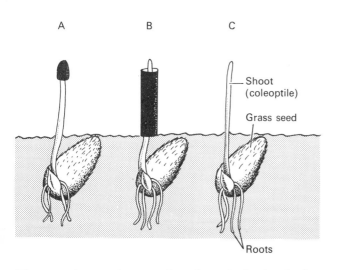

A   B   C

Shoot
(coleoptile)

Grass seed

Roots

Result after 24 hours

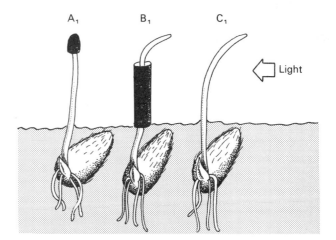

A₁   B₁   C₁

Light

Fig. 48.3   An experiment to show that only the tip of a shoot is sensitive to light

49

Geotropism and hydrotropism

Geotropism is growth movement in response to the pull of gravity.

If a bean seedling is grown in a horizontal position (Fig. 49.1A), its roots will curve downwards (Fig. 49.1B). This is called **positive geotropism** because the root grows in the *same* direction as the pull of gravity. But the shoot of this seedling will show **negative geotropism**, since shoots always grow upwards in the *opposite* direction to the pull of gravity.

If ink marks are drawn 1 mm apart on the root and shoot before the experiment, it is possible to see how the plant bends as it grows. Figure 49.1A and B show how the ink marks spread apart just behind the tips of the root and shoot.

To show that these growing regions are the only parts of a plant which can bend during tropic responses, simply turn the seedling through 90° (Fig. 49.1C). This brings the root and shoot back into a horizontal position. The curved parts do not straighten out. The plant produces new curves at the growing points behind the root and shoot tips so that the root continues to grow downwards and the shoot upwards (Fig. 49.1D).

As a control experiment a bean seedling can be grown in a horizontal clinostat. Place the clinostat in the dark and set it to rotate about once an hour (Fig. 49.2). The force of gravity acts equally on all parts of the plant, and the seedling does not remain in one position long enough to make a geotropic response. Its root and shoot therefore grow horizontally.

Figure 49.3 shows how to demonstrate **hydrotropism**, which means growth movement in response to the presence of water. In Figure 49.3B pea seedling roots are growing upwards (in the opposite direction to the pull of gravity) towards water in the bulb fibre. This is an example of **positive hydrotropism**.

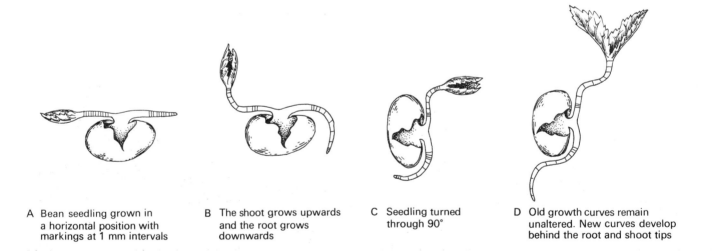

A Bean seedling grown in a horizontal position with markings at 1 mm intervals

B The shoot grows upwards and the root grows downwards

C Seedling turned through 90°

D Old growth curves remain unaltered. New curves develop behind the root and shoot tips

Fig. 49.1   How to demonstrate geotropism in roots and shoots

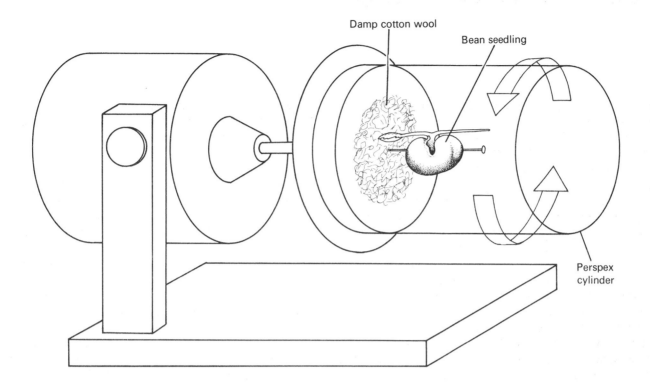

Damp cotton wool

Bean seedling

Perspex cylinder

Fig. 49.2 A clinostat (A bean seedling is rotated in the dark.
Its root and shoot grow horizontally.)

A Germinating peas in sieve

B A few days later

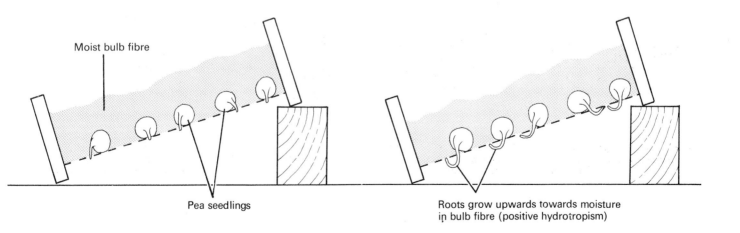

Moist bulb fibre

Pea seedlings

Roots grow upwards towards moisture
in bulb fibre (positive hydrotropism)

Fig. 49.3 How to demonstrate hydrotropism

50

The nervous system

The nervous system co-ordinates the workings of the body. This means that it controls body organs so that they work together at the correct times and at the correct rates according to the needs of the body.

The nervous system in humans, and in all other vertebrates, consists of a **brain** and a **spinal cord**. Together these form the **central nervous system** (Fig. 50.1). This system is connected to all parts of the body by **nerves**, which are made up of thousands of long, thin **nerve fibres**.

The nervous system co-ordinates the body by means of 'messages' called **nerve impulses**. These impulses travel to and from all parts of the body along nerve fibres.

The brain is the main co-ordinating centre of the body. The largest part of the human brain is the **cerebrum**, which consists of two **cerebral hemispheres** (Fig. 50.2A). The cerebrum receives infor-

mation from all the sense organs, and uses it to co-ordinate conscious behaviour: everything from eating food to complex activities like playing a musical instrument. The cerebrum also stores information (in other words, remembers it), and is responsible for mental activities such as decision making and solving mathematical problems.

Below the cerebrum is the **cerebellum**. This receives impulses from the organs of balance in the inner ear and from sense organs which detect movement and tension in muscles. This information is used to maintain balance and co-ordination of muscles during activities such as walking, running, dancing, or riding a bicycle.

Below the cerebellum is the **medulla oblongata**. This controls the rates of breathing and heart-beat, blood pressure, body temperature, and many other unconscious processes, which a person is not usually aware of, and cannot voluntarily control.

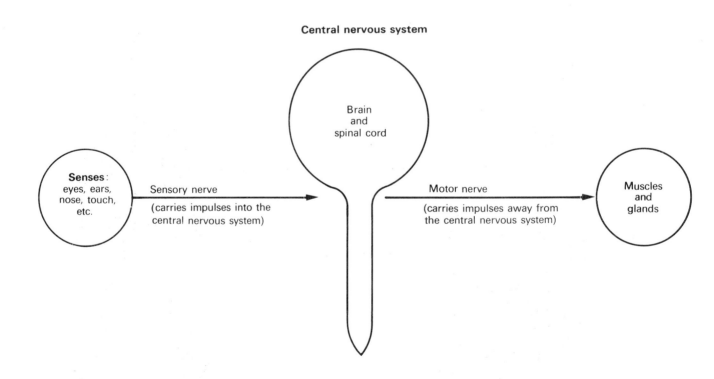

Central nervous system

Brain and spinal cord

Senses: eyes, ears, nose, touch, etc.

Sensory nerve (carries impulses into the central nervous system)

Motor nerve (carries impulses away from the central nervous system)

Muscles and glands

Fig. 50.1   Diagram of the nervous system

A The brain showing the specialized areas of the cerebrum

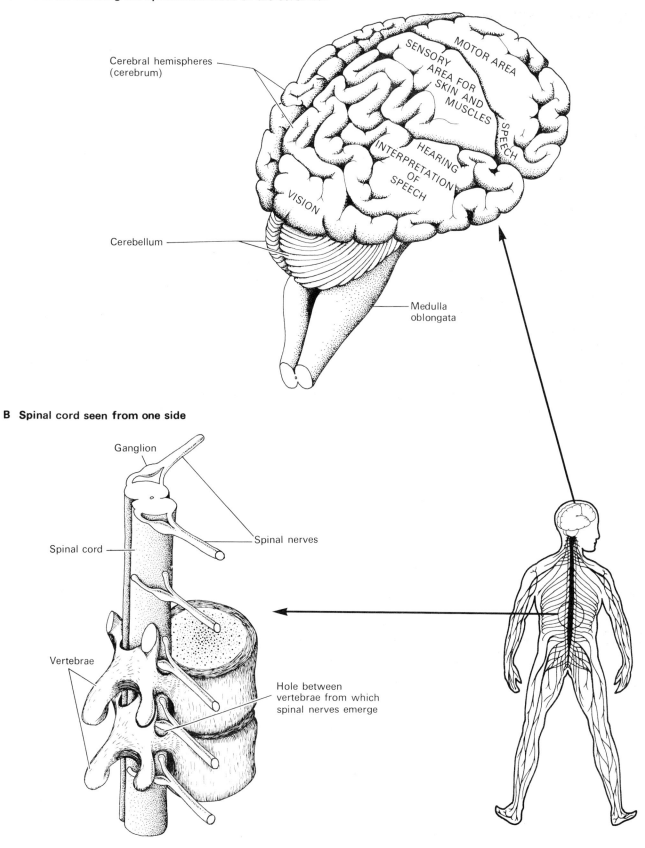

Cerebral hemispheres
(cerebrum)

MOTOR AREA

SENSORY AREA FOR SKIN AND MUSCLES

SPEECH

HEARING

INTERPRETATION OF SPEECH

VISION

Cerebellum

Medulla
oblongata

B Spinal cord seen from one side

Ganglion

Spinal nerves

Spinal cord

Vertebrae

Hole between
vertebrae from which
spinal nerves emerge

Fig. 50.2 The human nervous system

Nerve cells and reflexes

Nerve cells differ from other cells because their cytoplasm extends outwards from the main cell body, forming long, fine threads. These threads are the nerve fibres along which nerve impulses travel. Impulses travel along nerve fibres in only *one* direction. They travel into the central nervous system along fibres of **sensory nerve cells**, and out of the central nervous system along fibres of **motor nerve cells** (Figs. 50.1 and 51.1). The simplest way to show how nerve cells work is to describe a **reflex action**. During a reflex action a person responds to something automatically without having to think about it. For example, if the hand touches something hot it is pulled away quickly without conscious thought; if dust blows into the eyes they water and blinking occurs automatically until the dust is removed; and if a bright light suddenly shines into the eyes the pupils automatically contract.

The reflex action described in Figure 51.2 is a response to pain. Impulses from a stimulated pain receptor travel along a sensory nerve fibre to the spinal cord. First, the impulses travel through the **white matter** of the cord (mainly nerve fibres), and then they enter the **grey matter**. Here, impulses pass from the sensory fibre to another nerve cell across a tiny gap called a **synapse**. Only a relatively strong stimulation of a sense organ can cause sufficient impulses to cross this and other synapses. This is why weak stimulation of a sense organ has no effect on the nervous system. Impulses travel through the grey matter along a **relay nerve cell**. Then they cross another synapse and pass out of the spinal cord along the fibre of a motor nerve cell. The motor nerve cell carries the impulses to a muscle where they cause contraction. The contraction produces the response: the foot is removed from the painful stimulus. All this takes place in only a fraction of a second.

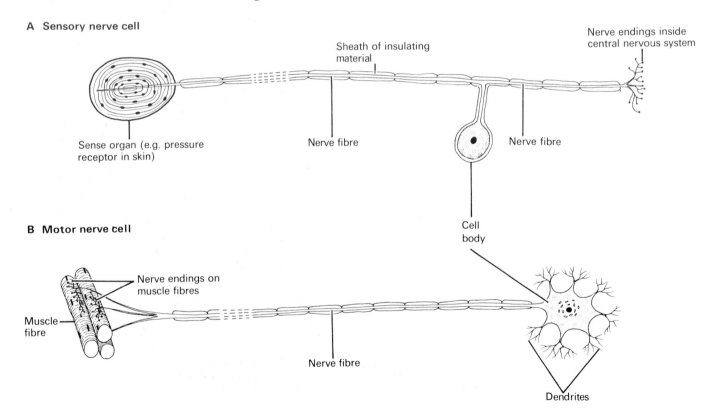

Fig. 51.1 Types of nerve cell

Stimulus and response

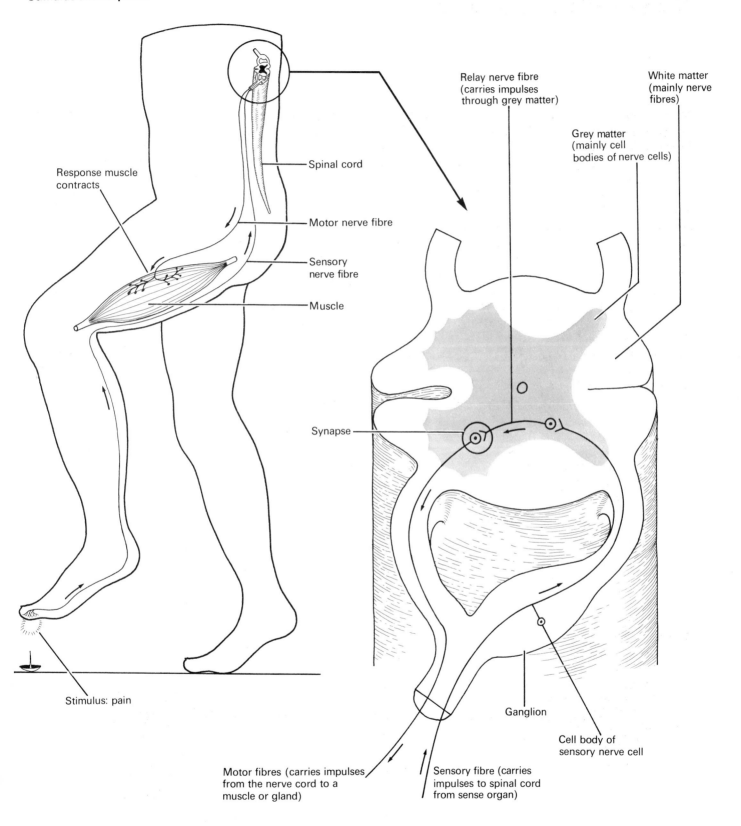

Response muscle contracts

Spinal cord

Motor nerve fibre

Sensory nerve fibre

Muscle

Relay nerve fibre (carries impulses through grey matter)

White matter (mainly nerve fibres)

Grey matter (mainly cell bodies of nerve cells)

Synapse

Stimulus: pain

Ganglion

Cell body of sensory nerve cell

Motor fibres (carries impulses from the nerve cord to a muscle or gland)

Sensory fibre (carries impulses to spinal cord from sense organ)

Fig. 51.2 An example of a reflex action

The endocrine system

The endocrine system consists of many different glands. These glands produce chemicals called **hormones.** Hormones are released into the blood-stream and are transported around the body. They have important effects on certain organs, and in some cases on the body as a whole.

Both hormones and nerve impulses co-ordinate the workings of the body, but they do this in different ways. The difference between them is rather like the difference between a telephone message and a message broadcast by radio. A telephone message goes along a wire to one person, and a nerve impulse goes along a nerve fibre to one particular muscle or gland. A radio message, however, is broadcast to everyone with a radio set, but only those actually concerned with the message respond to it. Similarly, hormones

are 'broadcast' by the blood-stream to every part of the body, but only certain parts respond to them (Fig. 52.1).

Some hormones, such as adrenalin, have temporary effects on the body. If a situation arises involving sudden emotional stress (fear or anger, for example), adrenalin causes an increase in heart-beat and breathing rates, and in the amount of glucose sugar in the blood. This prepares the body for action to cope with the situation. But when the emergency has passed these processes slow down again. Other hormones have long-lasting effects. Hormones from the pituitary gland, for example, affect the size to which the body grows. They also indirectly affect mental ability and the development of sexual characteristics.

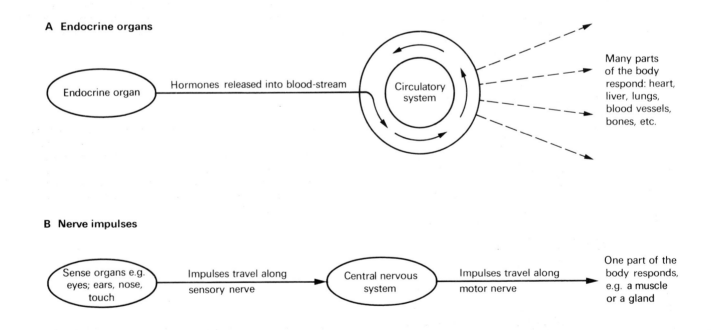

Fig. 52.1 Diagram showing how hormones and nerve impulses co-ordinate the body

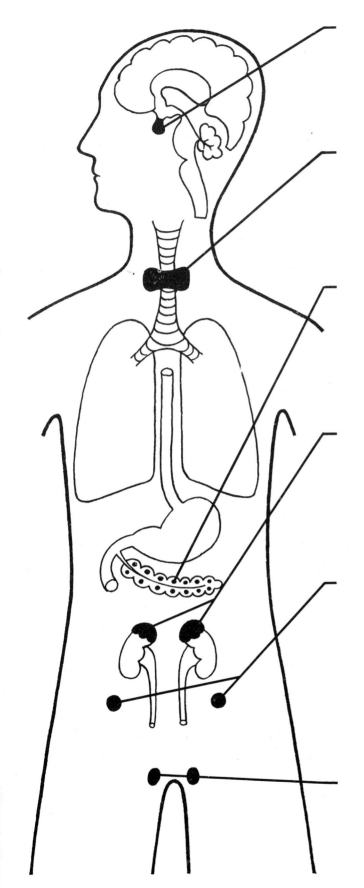

Pituitary   Situated at the base of the brain. Produces hormones which control growth. Too large an amount of these hormones causes **giantism**; too little causes **dwarfism**. Other hormones produced by the pituitary cause ovaries to release eggs; testes to produce sperms; the uterus to contract and expel the foetus at birth; and the mammary glands (breasts) to produce milk. Other hormones control the amount of water in urine; and the activities of the other endocrine glands.

Thyroid gland   Situated in the neck in front of the windpipe. Produces the hormone **thyroxine**, which has a major influence on physical and mental development after birth by controlling the rate of chemical reactions in all body cells. If too little thyroxine is produced in childhood **cretinism** develops (stunted physical and mental growth). In adults too little thyroxine causes overweight, thick skin and coarse hair, slow mental and physical reactions, and premature ageing. Too much thyroxine causes underweight, restlessness, and mental instability.

Pancreas   Situated below the stomach. One part produces digestive juice; other parts produce the hormone **insulin**, which decreases the rate at which the liver releases glucose into the blood; enables cells to absorb glucose; and stimulates the body to change glucose into fat. If too little insulin is produced the liver releases too much glucose. This causes the disease **sugar diabetes**, in which the level of glucose in the blood increases.

Adrenal glands   Situated on top of each kidney. Produce the hormone **adrenalin**. This prepares the body for action by raising blood pressure; increasing heart-beat and breathing rates; increasing the amount of glucose sugar released from the liver; and increasing the supply of blood to the muscles and reducing the supply to the gut.

Ovaries (females only)   Situated in the lower abdomen, below the kidneys. The parts of the female reproductive system that produce eggs. The ovaries also produce a sex hormone called **oestrogen**, which controls development of female secondary sexual characteristics (breasts, soft skin, feminine voice); prepares the uterus so that it can receive a fertilized egg; stimulates the uterus to protect and nourish a developing baby.

Testes (males only)   Situated in the groin, in a sac called the scrotum. The parts of the male reproductive system which produce sperms. Testes also produce a sex hormone called **testosterone**, which controls the development of male secondary sexual characteristics (deeper voice, coarser skin, and more body hair than in females).

Fig. 52.2   Endocrine organs and their functions

Asexual reproduction in Amoeba and yeast

Living things have the power to create new organisms by reproducing. In **asexual reproduction** only one parent is involved. The young are identical to their parent because they develop from only one set of hereditary characteristics. In contrast, **sexual reproduction** involves two parents. The young develop from two sets of hereditary characteristics, and are therefore identical to neither parent, though they may show some features of both parents (Fig. 53.1).

Amoeba

Like many other unicellular organisms, *Amoeba* reproduces asexually by the simplest of all methods: it divides into two parts. This is called **binary fission** (Fig. 53.2). A fully grown, well-fed amoeba becomes rounded in shape. Its nucleus divides into two and then its cytoplasm divides so that half of it surrounds each nucleus. The parent amoeba eventually splits into two daughter amoebas.

In cold weather or when food is in short supply, an amoeba may become rounded in shape and develop a hard protective covering called a **cyst**. In this **encysted** condition it can survive freezing, drought, and may be blown about in the air as dust. While it is encysted an amoeba divides many times by a process called **multiple fission**. When more favourable conditions return the cyst bursts open, releasing many new amoebas.

Yeast

Yeast is a unicellular fungus. It reproduces asexually by **budding**, a process in which the parent develops bud-like outgrowths (Fig. 53.3). A bud forms on the wall of the parent cell. At the same time its nucleus divides into two. One nucleus passes into the bud, and the other remains in the parent. The bud then either separates from the parent or remains attached, and may eventually produce another bud. If this continues a long chain of yeast cells may be formed.

Under favourable conditions budding occurs very fast, and each parent cell can produce thousands of daughter cells in a few days.

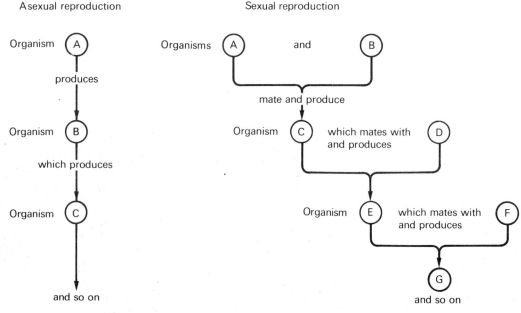

Young are identical to parents because the hereditary characteristics remain the same from one generation to the next

Young have features of both parents because two sets of hereditary characteristics are brought together each time fertilization occurs

Fig. 53.1 Diagram summarizing the main difference between sexual and asexual reproduction

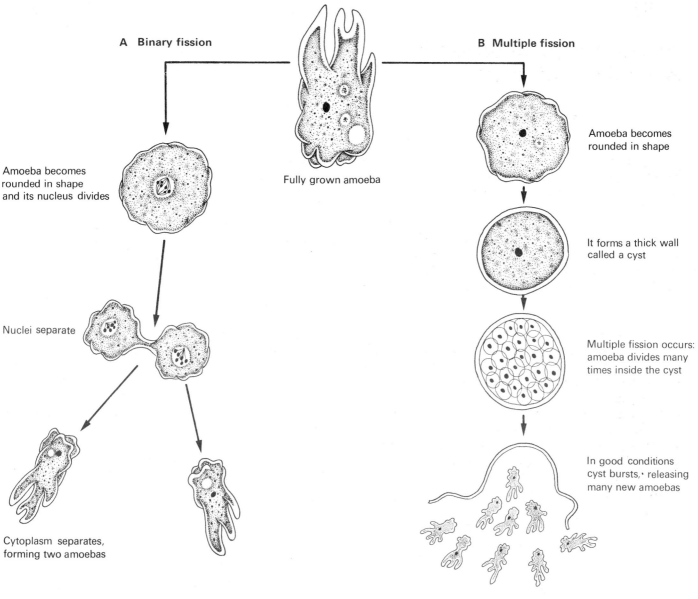

A Binary fission

B Multiple fission

Fully grown amoeba

Amoeba becomes
rounded in shape
and its nucleus divides

Amoeba becomes
rounded in shape

It forms a thick wall
called a cyst

Nuclei separate

Multiple fission occurs:
amoeba divides many
times inside the cyst

In good conditions
cyst bursts, releasing
many new amoebas

Cytoplasm separates,
forming two amoebas

Fig. 53.2   Asexual reproduction in *Amoeba*

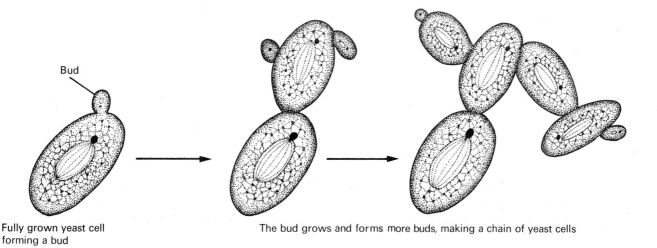

Bud

Fully grown yeast cell
forming a bud

The bud grows and forms more buds, making a chain of yeast cells

Fig. 53.3   Budding in yeast

The reproductive system of a female mammal

In mammals the young develop inside the body of the female, where they absorb food and oxygen from the mother's blood-stream, are kept at a constant warm temperature, and are protected from injury.

Eggs, or **ova**, are produced by two egg-shaped organs called **ovaries**. Ova are the female sex cells. They begin forming in the ovaries of human females before birth. At birth a baby girl has several hundred thousand partly formed ova in her ovaries. After birth no new ova are produced; in fact most of the existing ova disintegrate. The remaining ova continue growing until the girl becomes sexually mature: that is, until she reaches the age of **puberty**. This usually happens sometime between eight and fifteen years of age. At puberty the ova become fully developed one at a time, and are then released from

the ovaries by a process called **ovulation**. Ovulation occurs regularly about once a month in humans until about the age of fifty.

At ovulation an ovum is released from one of the ovaries into a **fallopian tube**. Each fallopian tube has a funnel-shaped opening and is lined with microscopic hairs called **cilia** (Fig. 54.1A). The cilia create a current of fluid which sweeps the ovum along inside the fallopian tube. Eventually the ovum reaches a wider, thicker-walled tube called the womb, or **uterus**. An ovum which has been fertilized by a sperm develops into a baby in the uterus.

A ring of muscle called the **cervix** closes the lower end of the uterus where it joins another tube, the **vagina**. The vagina extends to the outside of the body. The external opening is called the **vulva**.

A Front view

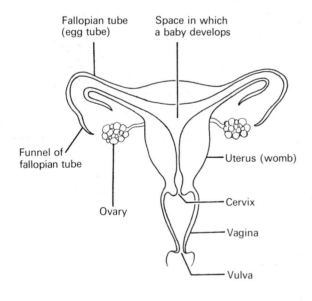

B Side view

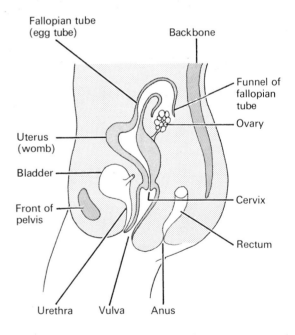

Fig. 54.1 Diagram of the reproductive system in a human female

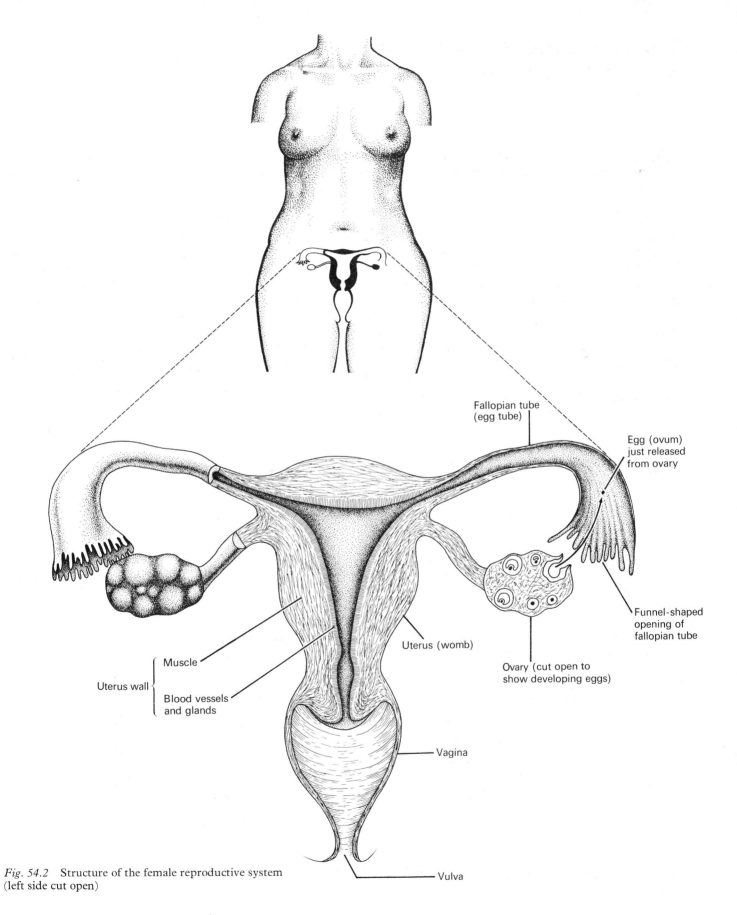

Fallopian tube
(egg tube)

Egg (ovum)
just released
from ovary

Funnel-shaped
opening of
fallopian tube

Uterus (womb)

Ovary (cut open to
show developing eggs)

Muscle

Uterus wall

Blood vessels
and glands

Vagina

Vulva

Fig. 54.2 Structure of the female reproductive system
(left side cut open)

Menstruation and the menstrual cycle

From the time a girl becomes sexually mature (reaches puberty) until she is about fifty years old she experiences a sequence of events called the **menstrual cycle**, which takes place about once every month. The various stages of the menstrual cycle are described below:

1. An ovum (egg) is released from one of her ovaries. This is called ovulation.

2. During and after ovulation the ovary produces a hormone called **oestrogen**, which causes the lining of the uterus to grow and develop a thick network of blood vessels. This prepares the uterus to receive and nourish the ovum if it is fertilized and begins to develop into a baby.

3. Meanwhile the ovum has been sucked into a fallopian tube and is travelling slowly along it towards the uterus. If the ovum is not fertilized within thirty-six hours after ovulation it dies.

4. About fourteen days after ovulation the ovary stops producing oestrogen, and a process called **menstruation** begins. The thick lining of the uterus breaks down and, together with the dead ovum, passes out of the body through the vagina. This discharge is called the **menstrual flow**.

Counting the beginning of menstruation as day 1, the menstrual flow usually stops on day 5. Ovulation usually occurs on day 14, and menstruation starts again on about day 28 (Fig. 55.2).

It is important to remember, however, that ovulation is not restricted to day 14 of the cycle. In most women ovulation can occur as early as day 13 or as late as day 15, and in some women ovulation can happen at almost any stage of the cycle. Immediately after ovulation a woman is said to be **fertile** and may **conceive**, or become **pregnant**, if she has sexual intercourse. It is almost impossible to be certain exactly when ovulation takes place, and sperms can live for up to three days in the female reproductive system. This means that most women must assume that they are fertile from at least day 11 to day 17 of their cycle (Fig. 55.1).

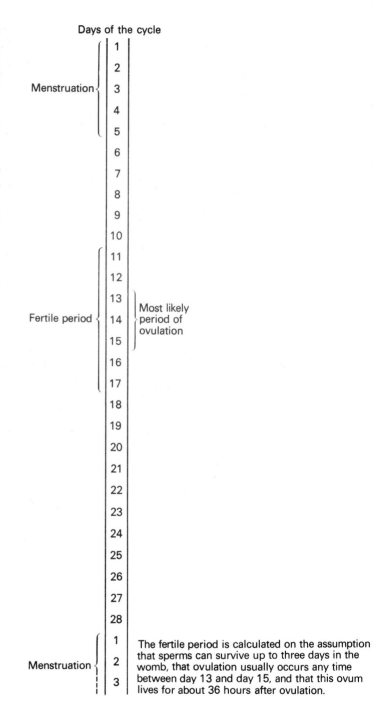

Days of the cycle

Menstruation { 1 2 3 4 5 }
6
7
8
9
10
Fertile period { 11 12 13 14 15 16 17 }

Most likely period of ovulation

18
19
20
21
22
23
24
25
26
27
28
Menstruation { 1 2 3 }

The fertile period is calculated on the assumption that sperms can survive up to three days in the womb, that ovulation usually occurs any time between day 13 and day 15, and that this ovum lives for about 36 hours after ovulation.

Fig. 55.1   Menstrual cycle in a human female

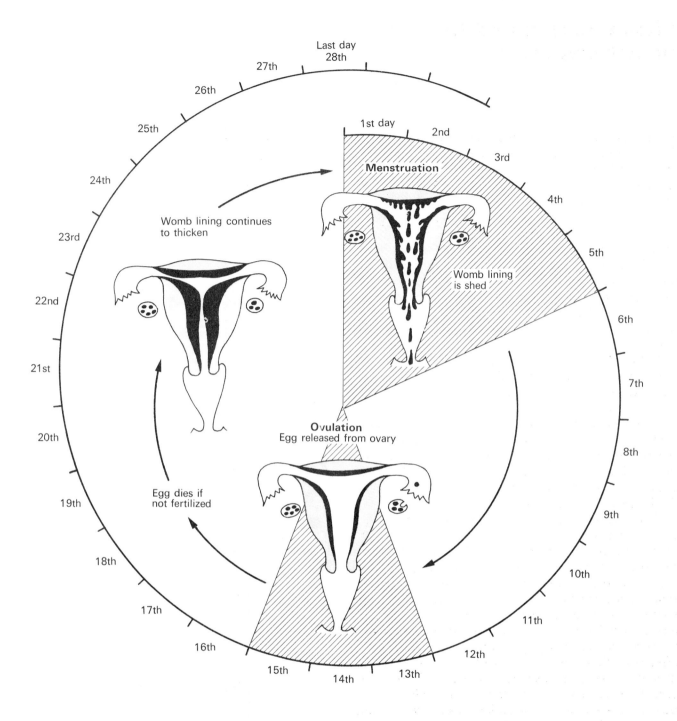

Last day
28th
27th
26th
25th
24th
23rd
22nd
21st
20th
19th
18th
17th
16th
15th
14th
13th
12th
11th
10th
9th
8th
7th
6th
5th
4th
3rd
2nd
1st day

Menstruation

Womb lining
is shed

Ovulation
Egg released from ovary

Womb lining continues
to thicken

Egg dies if
not fertilized

Fig. 55.2   Diagram of the menstrual cycle

The reproductive system of a male mammal

Male sex cells are called **sperms**, and are produced in organs called **testes**. Human testes are oval in shape. They are about 5 cm long, and are contained in a protective bag called the **scrotum** (Fig. 56.1). Unlike ova, sperms are not produced once a month, but continuously from puberty until about seventy years of age.

The inside of a testis is divided into about three hundred compartments, each containing three narrow twisted tubes (Fig. 56.2). These tubes are lined with dividing cells which produce sperms. The hundreds of sperm-producing tubes join together and form a smaller number of collecting ducts. Sperms move out of the testis through these ducts into a single, coiled tube called the **epididymis**. This is a storage area for sperms. Each epididymis leads into a **sperm duct**, or **vas deferens**. The sperm ducts, one from each testis, rise up the body until they are joined by a duct from a **seminal vesicle** (Fig. 56.3).

Finally, the sperm ducts join together near the base of the bladder, forming a single tube called the **urethra**. This junction occurs inside the **prostate gland**. The urethra leads to the outside of the body through the **penis**.

The walls of the penis contain sponge-like spaces called **erectile tissue**. When a male is sexually stimulated, the erectile tissue fills with blood, making the penis erect and firm. During sexual intercourse, or **copulation**, the erect penis is inserted into the vagina of the female and moved back and forth. These movements stimulate sense organs in the penis and eventually cause rhythmic contractions in the walls of the sperm ducts, which pump a liquid called **semen** into the female. This event is called an **ejaculation**. Semen consists of millions of sperms and a liquid from the seminal vesicles and prostate gland which causes the sperm tails to make swimming movements.

Male reproductive system (front view)

Ureter

Bladder

Epididymis

Seminal vesicle

Prostate gland

Sperm duct

Scrotum   Testis   Penis   Urethra

Fig. 56.1   Male reproductive system (front view)

Vertical section through a testis

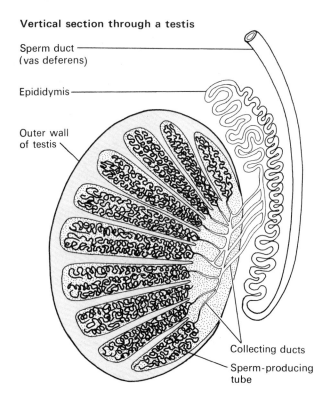

Sperm duct (vas deferens)

Epididymis

Outer wall of testis

Collecting ducts

Sperm-producing tube

Fig. 56.2   Testis cut in half

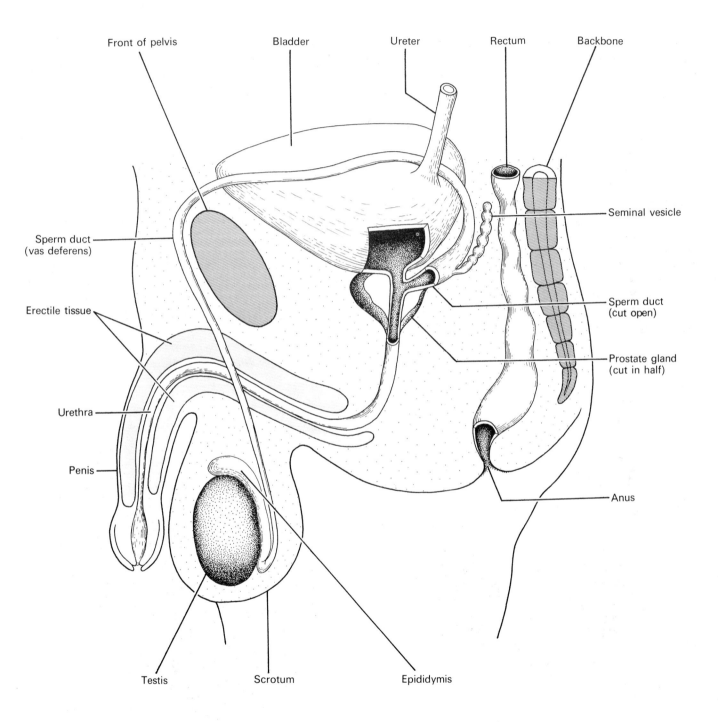

Front of pelvis

Bladder

Ureter

Rectum

Backbone

Sperm duct
(vas deferens)

Seminal vesicle

Erectile tissue

Sperm duct
(cut open)

Urethra

Prostate gland
(cut in half)

Penis

Anus

Testis

Scrotum

Epididymis

Fig. 56.3   Structure of the male reproductive system

57

Ovulation, fertilization, and implantation

Ovulation

The ovaries of a sexually mature human female contain ova (eggs) at various stages of development (Fig. 57.2). During the final stage of development an object called a **follicle** is produced. This consists of the ovum, and a mass of follicle cells which enclose a large bubble of liquid. A fully developed follicle bulges out from the surface of the ovary, and may measure up to 1 cm in diameter.

During the last few days of its development the follicle produces a hormone which causes the lining of the uterus to thicken, and the number of blood vessels in it to increase. The pressure in the follicle becomes so great that it bursts, shooting the ovum out of the ovary. This process is called ovulation. The ovum is then sucked into the funnel-shaped opening of one of the fallopian tubes. The remains of the follicle in the ovary collapse, but its cells continue growing and producing hormones which prepare the uterus lining for a fertilized ovum.

Fertilization

The semen ejaculated into the female during copulation contains up to 100 million sperms, but only one sperm can actually fertilize the ovum. Sperms swim up the uterus and into the fallopian tubes. The first sperm to reach the ovum penetrates its cell membrane and burrows into its cytoplasm (Fig. 57.1B). A thick **fertilization membrane** is formed. This is a skin which surrounds the ovum and prevents other sperms from entering (Fig. 57.1C). The sperm nucleus moves towards the ovum nucleus and the two fuse. This fusion brings together the hereditary characteristics of the mother and the father.

Implantation

The fertilized ovum takes about seven days to reach the uterus. During this time the ovum rapidly divides many times to form a hollow ball, containing hundreds of cells. This is called an **embryo**. The embryo comes to rest on the lining of the uterus, and produces digestive enzymes which dissolve the uterus lining. A hole is made, into which the embryo sinks (Fig. 57.2G and H). The embryo is now firmly embedded in the uterus wall. This process is called **implantation**.

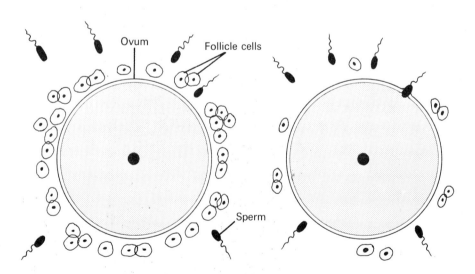 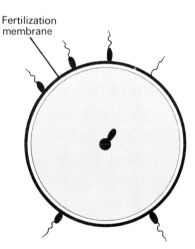

A   Sperms swim towards ovum

B   One sperm penetrates ovum

C   Fertilization membrane prevents entry of other sperms. Sperm and ovum nuclei fuse together

Fig. 57.1   Fertilization

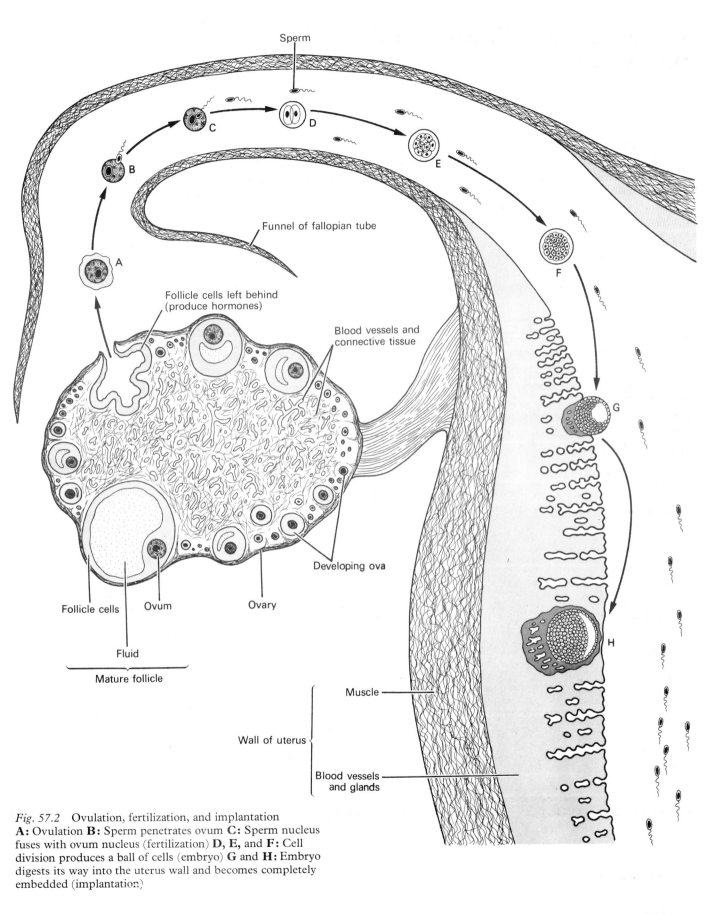

Sperm

C

D

B

E

Funnel of fallopian tube

A

F

Follicle cells left behind
(produce hormones)

Blood vessels and
connective tissue

G

Developing ova

Follicle cells

Ovum

Ovary

H

Fluid

Mature follicle

Muscle

Wall of uterus

Blood vessels
and glands

Fig. 57.2 Ovulation, fertilization, and implantation
A: Ovulation **B:** Sperm penetrates ovum **C:** Sperm nucleus
fuses with ovum nucleus (fertilization) **D, E,** and **F:** Cell
division produces a ball of cells (embryo) **G** and **H:** Embryo
digests its way into the uterus wall and becomes completely
embedded (implantation)

Pregnancy and the placenta

The period of development between fertilization and birth is called **pregnancy**. In humans it lasts about nine months. During this time hormones in the mother's blood prevent the monthly events of ovulation and menstruation.

The embryo must have oxygen and food in order to develop. At first it obtains these directly from its mother's blood as it flows through blood vessels in the lining of the uterus. Within three weeks of implantation, however, food and oxygen are absorbed within an organ called the **placenta**.

The placenta is a disc-shaped organ with millions of tiny root-like outgrowths called **villi**. These are embedded in the lining of the uterus. The placenta is connected to the embryo by a tube called the **umbilical cord**. The embryo's heart pumps blood through the umbilical cord into the placenta, where it flows through capillaries in the placental villi. The villi hang inside spaces in the uterus wall filled with the mother's blood. In these spaces food and oxygen pass from the mother's blood into the embryo's blood, while at the same time carbon dioxide and

other wastes pass from the embryo's blood into the mother's blood.

It is important to remember that the mother's blood does not flow *through* the embryo: it flows *past* villi containing the embryo's blood and it is here that food, oxygen, and waste are exchanged. In a sense the placenta takes the place of the baby's lungs, digestive system, and kidneys until birth, because obviously these organs cannot be used while the baby is inside the uterus.

The placenta is also an endocrine organ: that is, it produces hormones. These hormones ensure that the uterus grows at the same rate as the baby, and they stimulate glands in the breasts (mammary glands) to produce milk ready to feed the baby when it is born.

The baby develops inside a bag of liquid called the **amnion** (Figs. 58.1 and 58.3). The amnion acts as a shock-absorber and protects the baby from jolts and knocks as the mother moves about. It also helps to prevent the baby being injured if anything hits or presses against the mother's body.

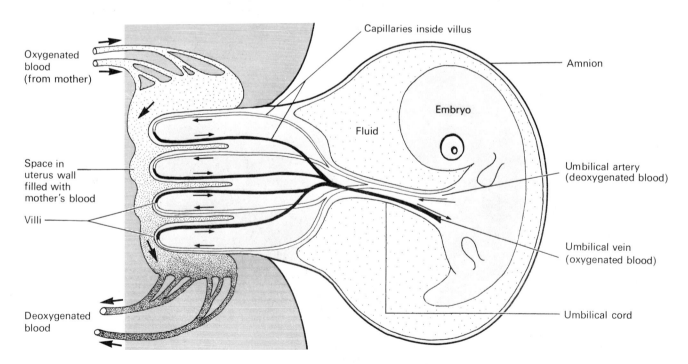

Fig. 58.1 Diagram of an embryo and its placenta

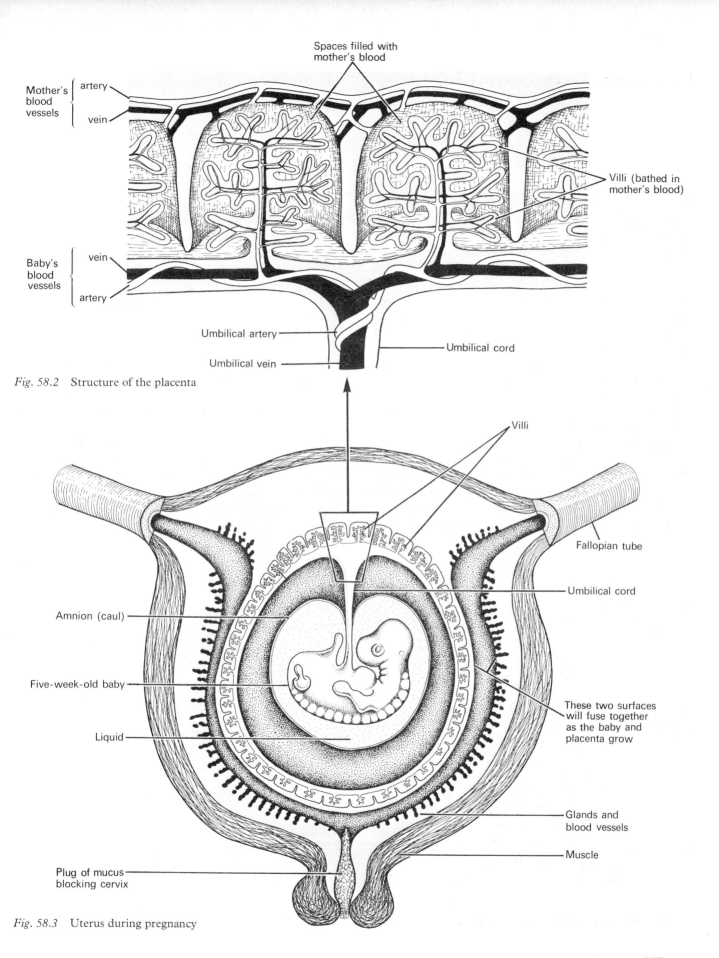

Spaces filled with
mother's blood

Mother's
blood
vessels {
artery

vein

Villi (bathed in
mother's blood)

Baby's
blood
vessels {
vein

artery

Umbilical artery

Umbilical cord

Umbilical vein

Fig. 58.2   Structure of the placenta

Villi

Fallopian tube

Umbilical cord

Amnion (caul)

Five-week-old baby

These two surfaces
will fuse together
as the baby and
placenta grow

Liquid

Glands and
blood vessels

Muscle

Plug of mucus
blocking cervix

Fig. 58.3   Uterus during pregnancy

117

Development and birth

A fertilized human ovum is less than the size of a full stop on this page. After five weeks it will have grown into an embryo 5 mm long, with a developing brain, eyes, and ears, a heart which pumps blood to and from its placenta, and tiny lumps on its sides which will soon be arms and legs. At six weeks it is 13 mm long, and its face and limbs are beginning to look human. After eight weeks it can no longer be called an embryo, because it is clearly human in appearance, with a full set of organs, and well developed limbs with fingers and toes. From this stage until birth it is called a **foetus**.

Towards the end of the seventh month the foetus is 35 cm long, has hair and eye-lashes, finger and toe nails, and milk teeth are developing in its jaws. Its internal organs are so well developed that it could remain alive if it were removed from its mother. In technical terms it is **viable**.

At nine months, just before birth, a baby weighs between 3 and 3·5 kg, and is about 50 cm long.

Birth is a dangerous time for a baby, because it is the moment when it leaves the protection of the uterus, emerges into the air, and is cut off from its supply of food and oxygen through the placenta.

During pregnancy the uterus wall develops thick muscles which contract and push the baby out of the mother's body when it is born. A few days before birth the baby turns round in the uterus until its head points downwards towards the cervix. The process of birth begins when the uterus walls start contracting rhythmically. The contractions are controlled by hormones in the mother's blood. They are weak at first, but gradually become more powerful, causing labour pains. The cervix opens and the baby's head passes into the vagina. This bursts the amnion and the fluid in it escapes. Further contractions push the baby right out of the mother's body.

Shortly after the baby is born more contractions push the placenta and umbilical cord out of the mother's uterus. These are called the **afterbirth**.

Position of baby immediately before birth

Baby emerges head first

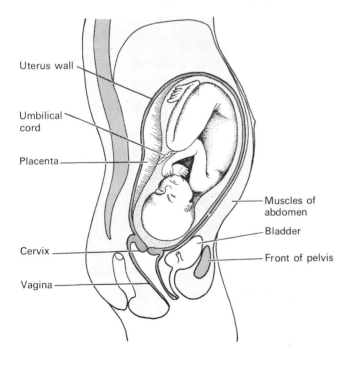

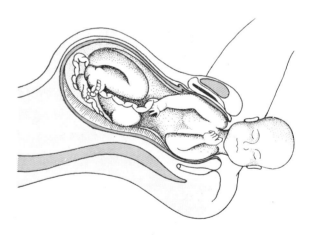

Uterus wall

Umbilical cord

Placenta

Cervix

Vagina

Muscles of abdomen

Bladder

Front of pelvis

Fig. 59.1 Birth

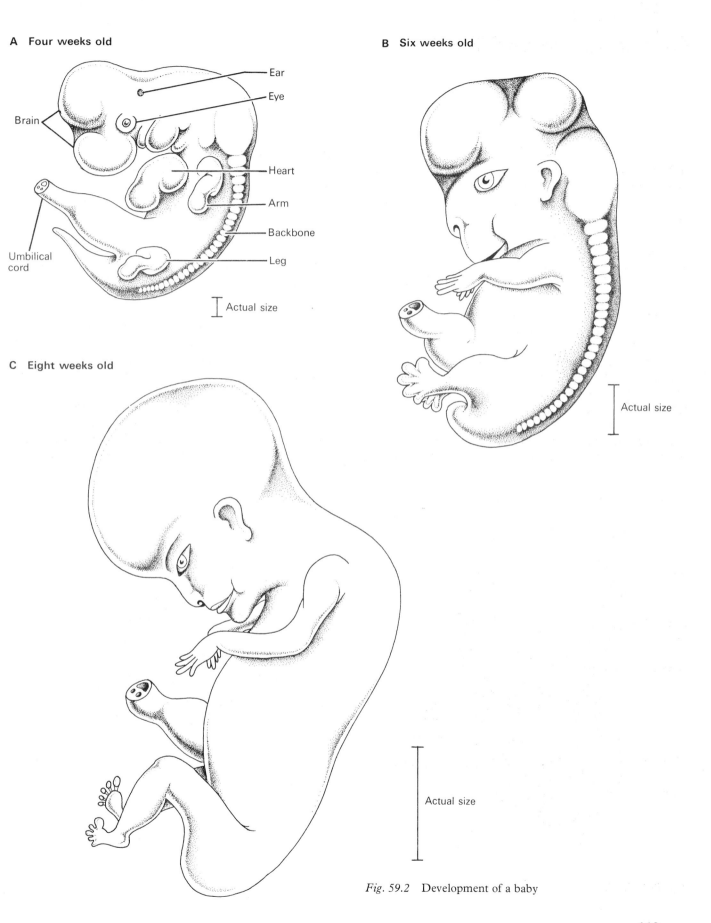

A Four weeks old

Ear

Eye

Brain

Heart

Arm

Backbone

Umbilical cord

Leg

Actual size

B Six weeks old

Actual size

C Eight weeks old

Actual size

Fig. 59.2 Development of a baby

119

Reproduction in birds

The main features of bird reproduction are: court-ship displays, nest-building, eggs with hard protective shells, and the care with which parents look after their young.

In Britain courtship behaviour in birds starts in spring. It includes special songs, movements of the body, and displays of feathers, most of which are performed by male birds. This behaviour enables a female to recognize males of her own species, and it stimulates her so that she is willing to copulate.

Both male and female birds have an opening between their legs called a **cloaca**. During copulation the male presses his cloaca against the female's and ejaculates sperms into a tube called an **oviduct** inside the female. Sperms swim up the oviduct and fertilize the newly formed eggs, which at this stage have no shells. After fertilization the eggs pass down the oviduct. On the way they are first enclosed in a clear jelly called **albumen** (the 'white' of the egg) and then covered by a shell. They are eventually laid, pointed end first, into the nest.

The nest is made by one or both parents and may be anything from a simple hollow scratched in the ground to an elaborate structure made from twigs, straw, and feathers. Usually the female **incubates** the eggs. This means she sits on them to keep them warm. She may lose some of her breast feathers during the nesting season. This exposes a warm area of skin called the **brood patch** which she presses against the eggs.

Most birds continue to sit on the nest long after the eggs have hatched, keeping the nestlings warm until their feathers grow. The young are fed by one or both parents, usually on caterpillars, grubs, worms, and insects. Parents protect their young when danger threatens. They may attack intruders, or lead them away from the nest.

When their feathers have developed, young birds take short practice flights around the nest. Until their strength and flying ability improve they are easily caught by cats, owls, hawks, etc. Only the strongest survive to begin breeding in the following year.

Reed buntings begging for food

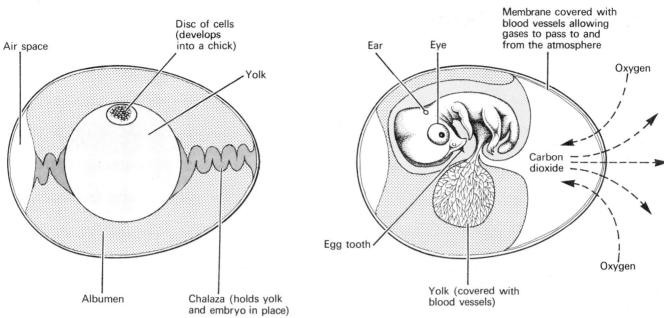

Air space

Disc of cells (develops into a chick)

Yolk

Albumen

Chalaza (holds yolk and embryo in place)

Ear

Eye

Membrane covered with blood vessels allowing gases to pass to and from the atmosphere

Oxygen

Carbon dioxide

Oxygen

Egg tooth

Yolk (covered with blood vessels)

Fig. 60.1 Development of a chicken

A chick hatching out

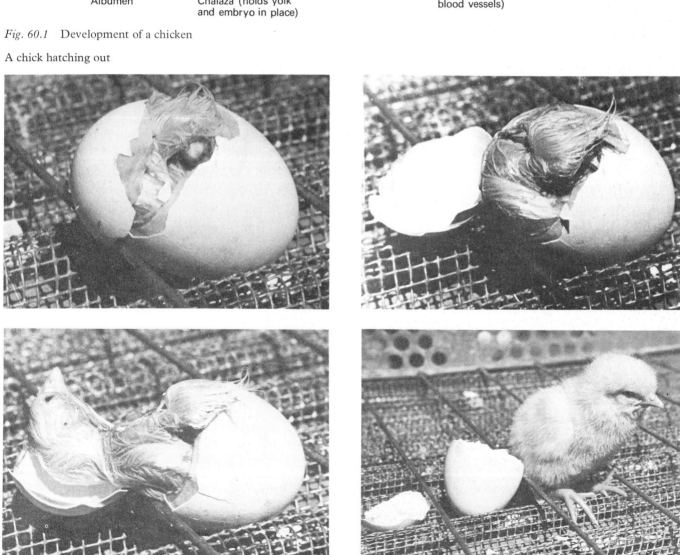

Reproduction in frogs

Frogs can live on land, but they must return to water to lay and fertilize their eggs. Their eggs develop into a stage called a **larva**, which is very different from the adult. A frog larva is called a tadpole, and is more like a fish than a frog. The change from larva to adult is called **metamorphosis**.

A female swollen with eggs is mounted by a male. He climbs on her back and holds her by the thick pads on his thumbs. The female lays about a thousand eggs, which are fertilized by sperms from the male as soon as they emerge from her body.

In the ten days following fertilization the egg becomes first an **embryo** and then a tadpole, which struggles out of the albumen and attaches itself to the outside of the albumen by its sticky **mucous gland** (Fig. 61.1A, B, and C). The tadpole's mouth is not yet open, and it still feeds on the yolk supply in its intestine. It breathes with external gills. Soon the tadpole goes swimming among water weeds, scraping microscopic plants from their leaves with the hard horny lips of its newly opened mouth.

Three weeks after hatching the external gills shrivel and are replaced by internal gills like those of a fish. A fold of skin, the **operculum**, grows over the new gills, and water which has passed over them now leaves the body through a hole called the **spiracle** (Fig. 61.1D and E).

Hind legs begin to develop after about two months, and front legs develop inside the operculum. Lungs develop and the tadpole has to swim regularly to the

Frogs mating

surface of the water to breathe air. The rasping lips disappear, and strong jaws develop which the tadpole uses to catch and eat small water animals.

Metamorphosis into an adult begins about three months after hatching. The tail is absorbed into the body; the legs become strong enough for swimming, and for jumping about on land; and a long sticky tongue develops, which the frog uses to catch insects. A newly developed frog is only 2 cm long (Fig. 61.1G). It must grow for another four years before it is old enough to begin breeding.

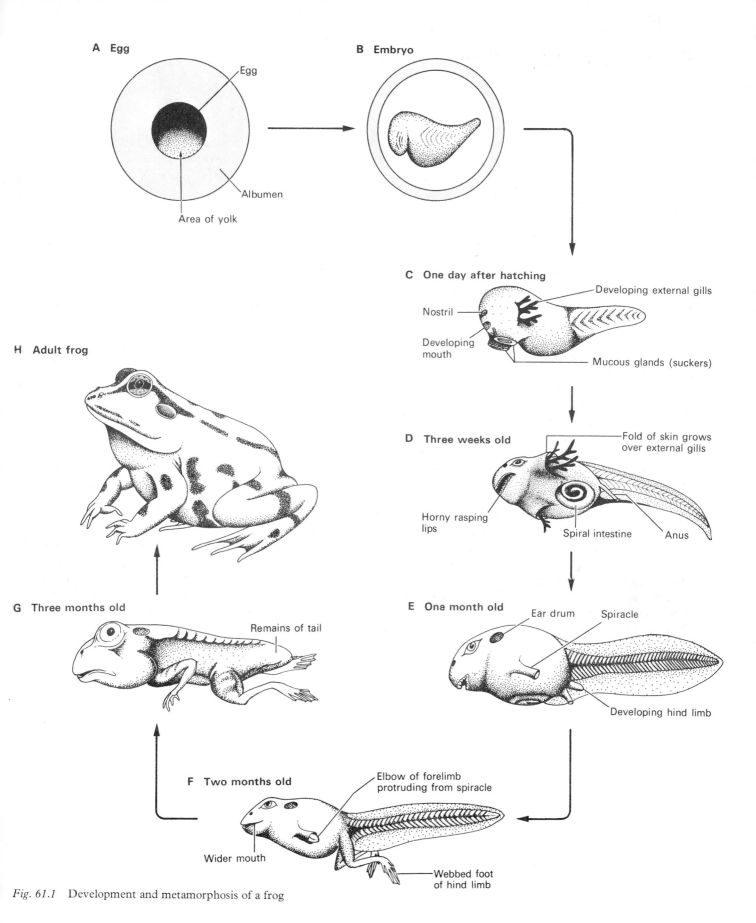

A Egg

Egg

Albumen

Area of yolk

B Embryo

C One day after hatching

Nostril

Developing mouth

Developing external gills

Mucous glands (suckers)

H Adult frog

D Three weeks old

Fold of skin grows over external gills

Horny rasping lips

Spiral intestine

Anus

G Three months old

Remains of tail

E One month old

Ear drum

Spiracle

Developing hind limb

F Two months old

Elbow of forelimb protruding from spiracle

Wider mouth

Webbed foot of hind limb

Fig. 61.1 Development and metamorphosis of a frog

Reproduction in fish (stickleback)

Fish reproduce in many different ways. The stickle-back is not typical of many other fish but it is common in British streams, and its reproductive behaviour can easily be observed in an aquarium.

In early spring male three-spined sticklebacks move to shallow water and begin nest building. During this time males develop bright blue eyes and orange-red undersides. The male digs a shallow hole by fanning the stream bed with his tail, and by scraping up mud in his mouth and spitting it out some distance away. He builds a roof over the hollow with vegetation stuck together with glue from his kidneys. The finished nest is a small tunnel open at both ends (Fig. 62.1). Other fish, especially other male sticklebacks, are driven away from the nest.

Any female stickleback whose body is swollen with eggs is approached by the male and courted. Other females are chased away. The male swims towards the female with quick zig-zag movements (Fig. 62.1). If the female notices him she responds by swimming towards him with her nose pointed upwards. The male turns and rushes off in the direction of his nest, and when he arrives he stops with his nose pointed at the entrance. If she does not respond by entering the nest, the male thrusts his nose into the entrance and turns on his side with his dorsal fin pointed towards her. If the female is ready to lay her eggs she will push her way into the nest until her nose protrudes from one end and her tail from the other. The male prods her belly with his nose and she lays her eggs. This is all over in a few seconds, and the female leaves the nest. She is immediately

Male three-spined stickleback carrying material to build nest

replaced by the male, who sheds sperms over the eggs, and fertilizes them. This process may be repeated several times with different females.

The male rears the young. He guards the eggs, and later the newly hatched young, which are called **fry**, against all intruders. He also spends long periods fanning the nest with his tail. This produces a current of water which carries oxygen to the eggs or fry.

The eggs hatch about ten days after fertilization, and the fry are kept in or near the nest by the male for at least three weeks. After this the male relaxes his guard and the fry swim away. By this time the male has lost most of his vivid colours.

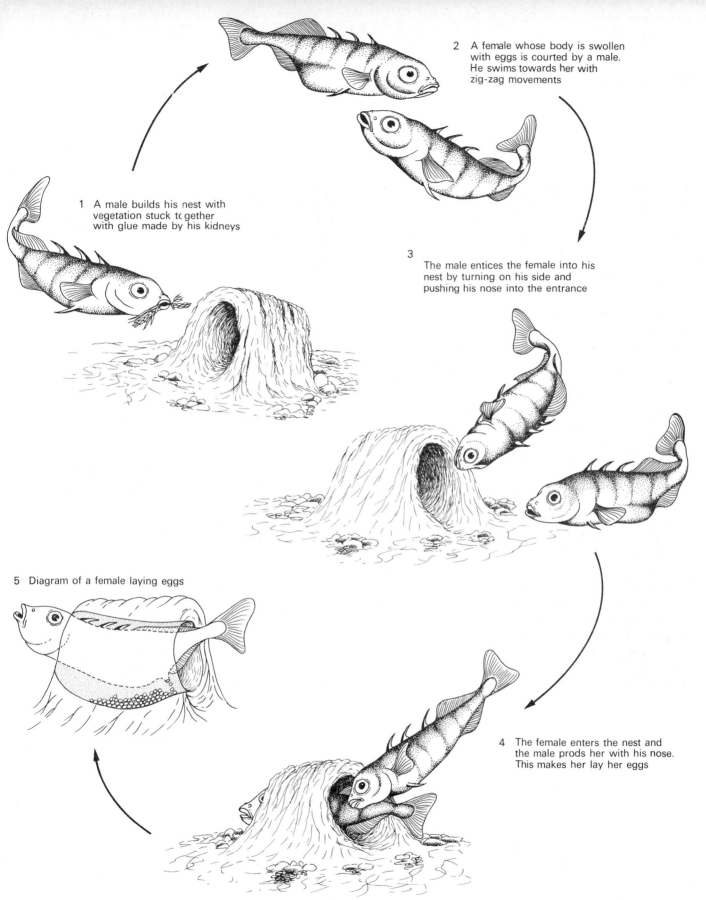

2 A female whose body is swollen with eggs is courted by a male. He swims towards her with zig-zag movements

1 A male builds his nest with vegetation stuck together with glue made by his kidneys

3 The male entices the female into his nest by turning on his side and pushing his nose into the entrance

5 Diagram of a female laying eggs

4 The female enters the nest and the male prods her with his nose. This makes her lay her eggs

Fig. 62.1 Nest-building, courtship, and egg-laying in three-spined sticklebacks

Reproduction in insects

In the group of insects which includes dragonflies, locusts, and cockroaches, the newly hatched young are called **nymphs**. A nymph looks very similar to its parents, but it is smaller, wingless, and unable to reproduce. Nymphs eat, grow, and shed their cuticles as they grow into adults. This pattern of development is called **incomplete metamorphosis**.

After mating, a female cockroach produces an egg case 10 mm long containing sixteen eggs (Fig. 63.1). She carries the egg case in her abdomen until she finds a warm dry place near food in which to lay it. The nymphs, like the adults, can eat many things: paper, leather, seeds, and discarded scraps of human food. Nymphs take three years to grow into winged adult cockroaches with reproductive organs. During this period they shed their cuticles six times.

In the group of insects which includes flies, bees, moths, and butterflies, the newly hatched young are called **larvae**. A larva looks very different from its parents, and its development into an adult is called **complete metamorphosis**.

After mating, a female cabbage white butterfly lays batches of about 100 eggs on the undersides of cabbage leaves. The eggs soon hatch into larvae called caterpillars (Fig. 63.2A and B). Caterpillars increase greatly in size, shedding their skins several times, but their cells do not divide. They simply inflate with digested food. About a month after hatching the caterpillars stop feeding and crawl to a dry sheltered spot. Here they first spin a pad of silk to which they attach their claspers, and then they spin a girdle of silk which holds the body in place (Fig. 63.2C). The bloated larval cells die, giving up their stored food to smaller cells which develop into a **pupa** (Fig. 63.2D). The common name for a butterfly pupa is a chrysalis. A pupa contains cells which produce an adult butterfly.

After fourteen days the pupa splits open and a soft, helpless butterfly emerges. At first its wings are crumpled and folded, but blood is pumped into them, making them expand to their full size. About an hour later the wings are hard and dry enough to enable the insect to fly. The scientific name for an adult insect is an **imago**.

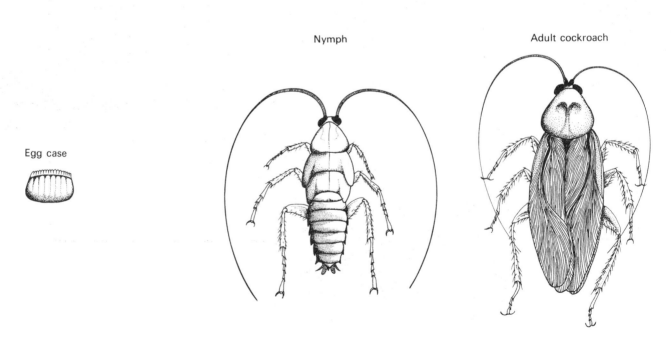

Egg case   Nymph   Adult cockroach

Fig. 63.1 Life history of a cockroach (incomplete metamorphosis)

126

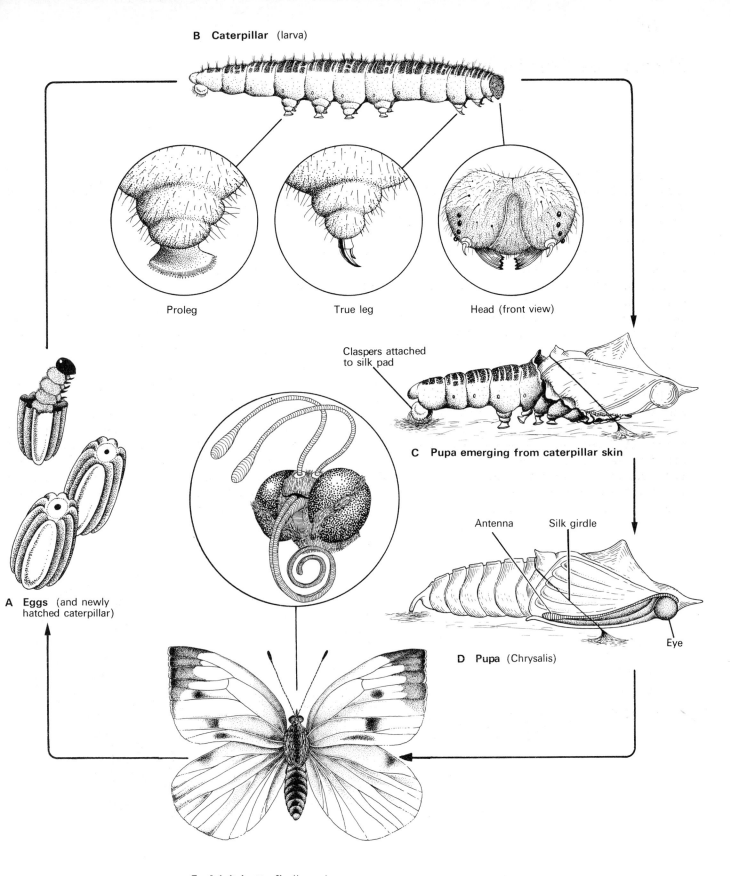

B Caterpillar (larva)

Proleg

True leg

Head (front view)

Claspers attached
to silk pad

C Pupa emerging from caterpillar skin

Antenna

Silk girdle

Eye

D Pupa (Chrysalis)

A Eggs (and newly
hatched caterpillar)

E Adult butterfly (imago)

Fig. 63.2 Life cycle of the cabbage white butterfly (**complete metamorphosis**)

Parts of a flower

The function of a flower is to carry out sexual reproduction. The reproductive organs produce fruits and seeds which develop into the next generation of plants.

The female reproductive organs, or **carpels**, are in the centre of a flower (Fig. 64.1C). A carpel consists of a hollow base or **ovary**, and a narrow **style** which ends in a **stigma**. During pollination the stigma becomes covered in pollen grains, either from the same flower or from another flower of the same species. The ovary contains one or more **ovules**. An ovule contains a large cell called the **embryo sac**. This cell is the female sex cell and its nucleus, the **egg nucleus**, is fertilized by a male nucleus from a pollen grain. The ovule then develops into a seed and the ovary wall becomes a fruit which contains the seeds. Some plants have separate carpels (e.g. buttercup), but most have a number of carpels fused together (e.g. poppy).

The male reproductive organs, or **stamens** (Fig. 64.1B), are arranged in a ring around the carpels. A stamen consists of a stalk called the **filament**. This ends in an **anther** made up of four **pollen sacs** in which pollen grains grow. Pollen grains are the male sex cells.

All flowers have the parts described above. Some or all of the following parts may also be present, depending how the flower is pollinated.

In most flowers the reproductive organs are surrounded by a ring of **petals**. Together these are known as the **corolla** of the flower. In some flowers the petals are brightly coloured and scented, and may have a **nectary** at the base which produces a liquid called **nectar**. These flowers are pollinated by insects which are attracted by the scent or the coloured petals, and come in search of nectar. As they pass from flower to flower they transfer pollen from one plant to another.

In many flowers the petals are surrounded by a ring of **sepals**. Together these are called the **calyx** of the flower. Sepals are usually green and form a protective outer covering when the flower is in bud.

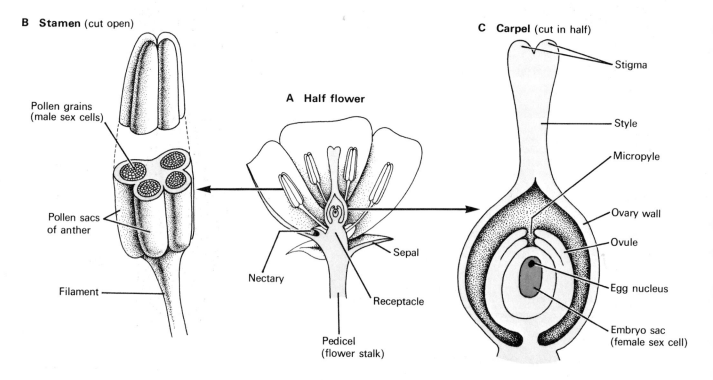

B Stamen (cut open)

Pollen grains (male sex cells)

Pollen sacs of anther

Filament

A Half flower

Nectary

Pedicel (flower stalk)

Sepal

Receptacle

C Carpel (cut in half)

Stigma

Style

Micropyle

Ovary wall

Ovule

Egg nucleus

Embryo sac (female sex cell)

Fig. 64.1 Parts of a flower (this flower is imaginary)

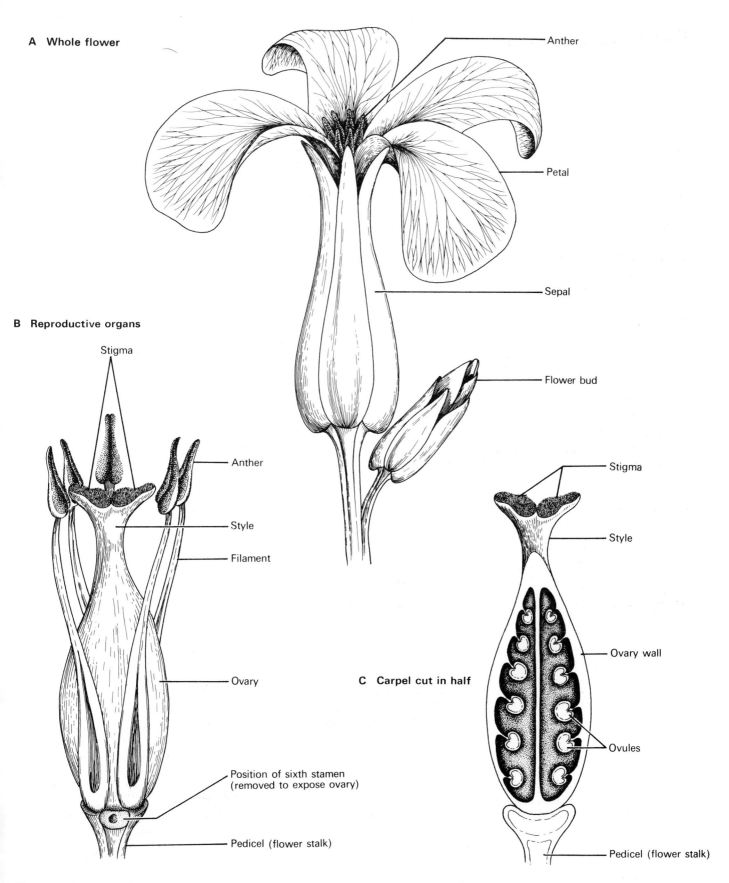

A   Whole flower

Anther

Petal

Sepal

Flower bud

B   Reproductive organs

Stigma

Anther

Style

Filament

Ovary

C   Carpel cut in half

Stigma

Style

Ovary wall

Ovules

Position of sixth stamen
(removed to expose ovary)

Pedicel (flower stalk)

Pedicel (flower stalk)

Fig. 64.2   Structure of a wallflower

129

Pollination

Pollination is the transfer of pollen grains from anthers to stigmas. **Self-pollination** is the transfer of pollen from anthers to stigmas in the same flower, or between flowers on the same plant. **Cross-pollination** is the transfer of pollen from one plant to another of the same species.

Many flowers have features which help reduce the chances of self-pollination, and increase the chances of cross-pollination. Obviously self-pollination is impossible where a plant has flowers of only one sex, that is, containing either stamens or carpels but not both. These are called **unisexual** flowers. In **bisexual flowers**, which have both stamens and carpels, self-pollination is prevented in two main ways. In plantains and white dead nettles, for example (Figs. 65.1 and 65.2), the carpels ripen and are pollinated before stamens in the same flower are ripe. In dandelions, however, the stamens ripen and release pollen before the carpels are ripe. But if cross-pollination does not take place, self-pollination is still possible (Fig. 65.3).

Pollen is usually transferred by wind or by insects. **Wind-pollinated** flowers include plantains (Fig. 65.1), grass, stinging nettles, and willows. Wind-pollinated flowers have no petals, or very small ones which are usually either white or green; they produce large numbers of tiny pollen grains which are easily carried by the wind; their large anthers have long filaments and hang outside the flower where the wind can catch them; they have spreading, feathery stigmas which act like a net, and catch pollen in the air; and their flowers are on long stalks, or develop before the leaves, so are not sheltered from the wind.

Insect-pollinated flowers include white dead-nettles, dandelions (Figs. 65.2 and 65.3), buttercups, and sweet peas. These flowers attract insects with their large, often brightly coloured petals, which may also be scented and have nectaries. They produce large pollen grains with spiky hairs which cling to insects' bodies. Their anthers and stigmas are situated inside the flower where they brush against insects in search of nectar (Fig. 65.2B and C).

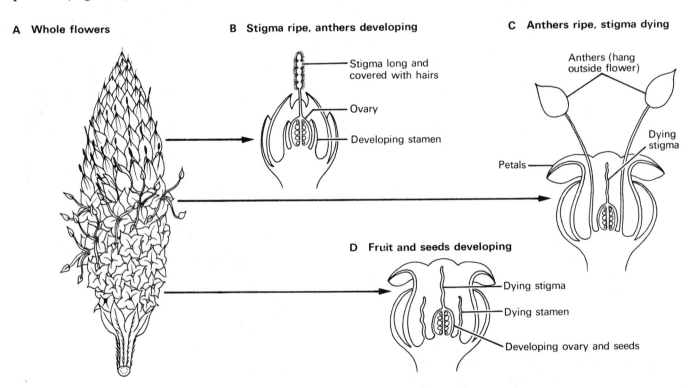

A Whole flowers

B Stigma ripe, anthers developing
- Stigma long and covered with hairs
- Ovary
- Developing stamen

C Anthers ripe, stigma dying
- Anthers (hang outside flower)
- Dying stigma
- Petals

D Fruit and seeds developing
- Dying stigma
- Dying stamen
- Developing ovary and seeds

Fig. 65.1   Wind pollination (plantain)

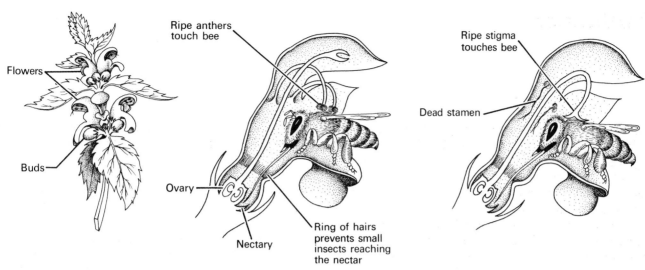

A Whole flowers

Flowers

Buds

B Half flower with ripe anthers

Ripe anthers
touch bee

Ovary

Nectary

Ring of hairs
prevents small
insects reaching
the nectar

C Half flower with ripe stigma

Ripe stigma
touches bee

Dead stamen

Fig. 65.2 Insect pollination (white dead nettle)

**A A closed stigma grows
upwards, pushing pollen
grains between the anthers**

**B Stigma opens into the
position for cross-pollination
by insects**

**C Stigma curves over into the
position for self-pollination**

One floret
(cut in half)

Stigma

Five anthers form a
tube around style

Style

Five petals
fused together

Nectary

Calyx (forms the
parachute of seed)

Receptacle

Ovary

Ovule

Fig. 65.3 Insect or self-pollination (dandelion). A dandelion is
a composite flower (made up of many florets). It is usually
pollinated by insects, but self-pollination is also possible: the
stigmas curve over until they touch pollen on their own styles

Fertilization and development of seeds and fruits

Fertilization is the fusing together of a **male gamete** and a **female gamete**. In flowering plants the male gamete is a nucleus inside a pollen grain, and the female gamete is the egg nucleus in the embryo sac of the ovule (Fig. 66.2D).

Fertilization occurs after pollen is transferred from the anthers of one flower to the stigmas of the same flower or another flower of the same type, or species. Fertilization does not occur if pollen is transferred between plants of different species.

After pollination the stigma produces a sugary liquid which feeds the pollen grains. Soon, a pollen grain bursts open and produces a long tubular outgrowth called a **pollen tube**. This tube grows down between the cells of the style, from which it receives more nourishment. The pollen tube enters the ovule through a tiny hole called the **micropyle**. It then grows into the embryo sac and its tip bursts open, forming a clear pathway through which the male gamete in the pollen grain can reach the female gamete. A nucleus from a cell in the pollen grain now moves down the pollen tube, enters the embryo

sac, and brings about fertilization by fusing with the egg nucleus.

After fertilization the stamens, petals, and eventually the sepals of the flower die and drop off. The ovules develop into seeds, and are enclosed in the ovary wall, which becomes the fruit (Fig. 66.1).

In **true fruits** the ovary wall becomes well developed. For instance, it forms the pods of peas and beans; the skin, flesh, and stone of a plum; the rind and flesh of an orange; and the shell of a nut. In **false fruits** the ovary wall remains undeveloped, and some other part of the flower becomes what is incorrectly called a 'fruit'. In strawberries, for example, the part normally called a fruit is really an enlarged fleshy receptacle. The true fruits are the tiny seed-like objects embedded in the flesh. In apples and pears the receptacle becomes thick and juicy and completely encloses the whole ovary, which is the core.

In both false and true fruits the development of the ovary wall or receptacle is usually associated with a mechanism which disperses the seeds some distance away from the parent plant.

A Pea flower cut in half

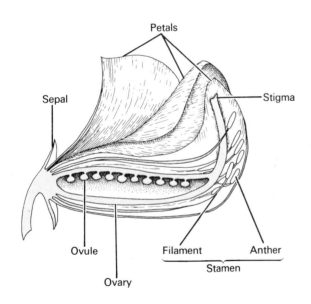

B Pea pods

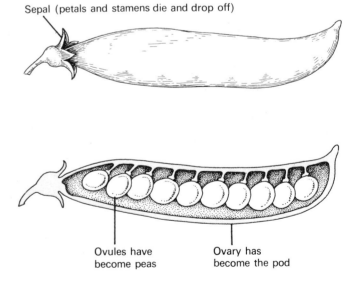

Fig. 66.1   Development of a pea flower after fertilization

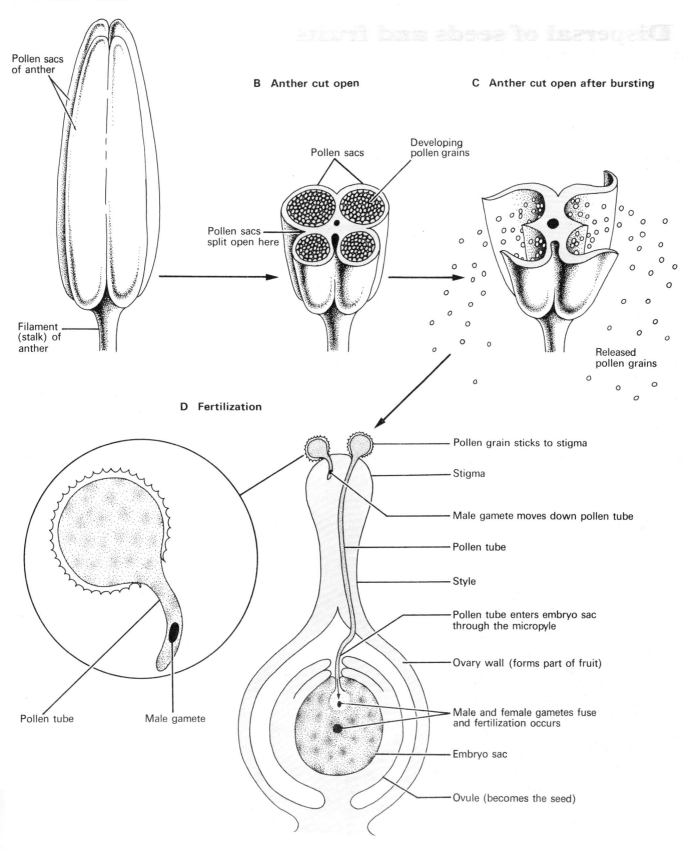

A Stamen intact

Pollen sacs
of anther

Filament
(stalk) of
anther

B Anther cut open

Pollen sacs

Developing
pollen grains

Pollen sacs
split open here

C Anther cut open after bursting

Released
pollen grains

D Fertilization

Pollen tube

Male gamete

Pollen grain sticks to stigma

Stigma

Male gamete moves down pollen tube

Pollen tube

Style

Pollen tube enters embryo sac
through the micropyle

Ovary wall (forms part of fruit)

Male and female gametes fuse
and fertilization occurs

Embryo sac

Ovule (becomes the seed)

Fig. 66.2 Diagram showing how pollen grains fertilize an ovule

Dispersal of seeds and fruits

A **dispersal mechanism** is something which carries fruits and seeds well away from the parent plant. These mechanisms help prevent overcrowding and often spread plants into new territory. Fruits and seeds are dispersed by wind, by animals, and sometimes by the parent plant itself.

Wind dispersal (*Fig. 67.1*)
In some plants the fruit or seed develops a large 'wing' or 'parachute' which is caught in the wind (e.g. sycamore and dandelion). In poppies and delphiniums the seeds are shaken out of the ovary as the plant sways in the wind.

Animal dispersal (*Fig. 67.2*)
Hooked fruits and seeds such as burdock and goosegrass may be carried for long distances attached to animals' fur. Succulent fruits and nuts attract animals as a source of food. Small hard-coated seeds such as blackberries, strawberries, and rose hips can pass through an animal's digestive system unharmed and may be carried some distance before being deposited on the ground with a convenient supply of fertilizer. In other plants, such as apple, plum, and cherry, the seeds in their hard coats are discarded after the soft fruit has been eaten by animals.

Self-dispersal (*Fig. 67.3*)
Several plants have mechanisms which throw seeds some distance from the parent. Most of these depend on tension caused by the drying of the fruit wall. The ripe pods of sweet pea, gorse, broom, and lupin suddenly split open and the two halves curl outwards, scattering the seeds. In the geranium, the styles curl up and out, throwing seeds from the cup-shaped ovaries. A similar mechanism occurs in the wallflower fruit, which splits open from the base upwards, scattering the seeds.

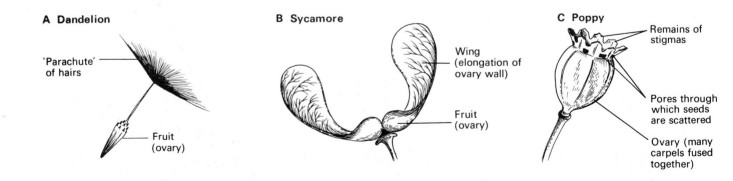

A Dandelion

'Parachute' of hairs

Fruit (ovary)

B Sycamore

Wing (elongation of ovary wall)

Fruit (ovary)

C Poppy

Remains of stigmas

Pores through which seeds are scattered

Ovary (many carpels fused together)

Fig. 67.1 Wind dispersal

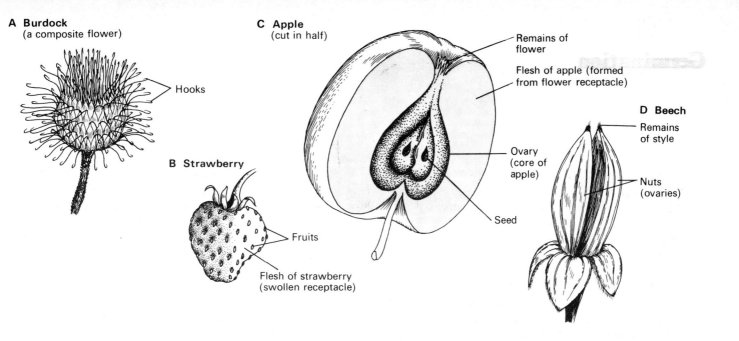

A Burdock
(a composite flower)

Hooks

C Apple
(cut in half)

Remains of flower

Flesh of apple (formed from flower receptacle)

D Beech

Remains of style

Nuts (ovaries)

B Strawberry

Fruits

Flesh of strawberry (swollen receptacle)

Ovary (core of apple)

Seed

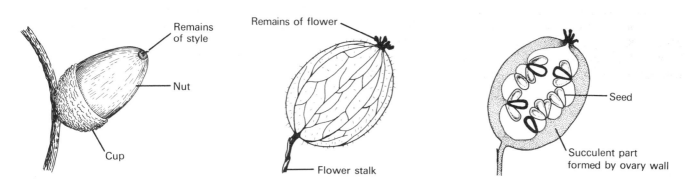

E Acorn

Remains of style

Nut

Cup

F Gooseberry (whole and cut in half)

Remains of flower

Flower stalk

Seed

Succulent part formed by ovary wall

Fig. 67.2 Animal dispersal

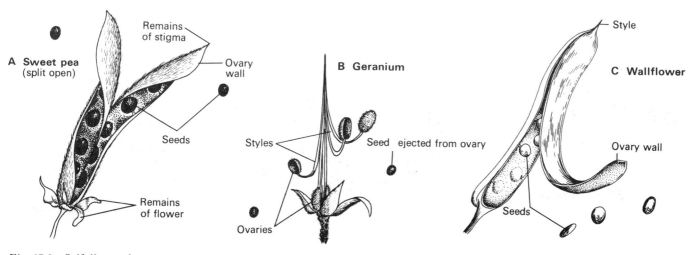

A Sweet pea
(split open)

Remains of stigma

Ovary wall

Seeds

Remains of flower

B Geranium

Styles

Seed ejected from ovary

Ovaries

C Wallflower

Style

Ovary wall

Seeds

Fig. 67.3 Self dispersal

Germination

A seed consists of a tiny embryo plant and a supply of stored food, all of which are enclosed in a protective covering called the seed-coat or **testa**. The embryo is made up of a young root, called the **radicle**; a young shoot, or **plumule**; and one or more seed leaves, or **cotyledons**. The stored food is found in one of two places. In peas and beans, for example (Fig. 68.1), the cotyledons are swollen with stored food. In cereals, such as barley, wheat, and maize (Fig. 68.2), the food is stored in a mass of dry powdery cells called **endosperm**. The flour used to make bread, cakes, etc., consists of endosperm. This is released when cereal grains are milled and the testa is removed.

Seeds appear to be dead; but if they are supplied with water, air, and warmth they 'come to life' and produce a new plant. This process is called **germination**.

A kidney bean, for example, absorbs water and swells (Fig. 68.3). The embryo plant begins to grow, using food stored in its cotyledons. First, the radicle bursts through the testa, grows into the soil, and produces root hairs through which it absorbs water and minerals. The cotyledons are then pulled up out of the soil in such a way that they protect the plumule sandwiched between them. The testa falls off. The plumule expands and produces the first foliage leaves, which begin photosynthesis. The cotyledons shrivel and drop off when their stored food is used up. Germination in which the cotyledons are brought above the ground is called **epigeal germination**.

The remains of a style on a maize grain show that it is actually a fruit and not a seed. After absorbing water, the grain produces a radicle, and this forms the root system. As the plumule grows up through the soil it is protected by a sheath called the **coleoptile**. When it is above the soil, the plumule bursts out of the coleoptile and expands, forming leaves. The single cotyledon remains below the soil absorbing food from the endosperm cells until photosynthesis makes the plant independent. Germination in which the cotyledons remain below ground is called **hypogeal germination**.

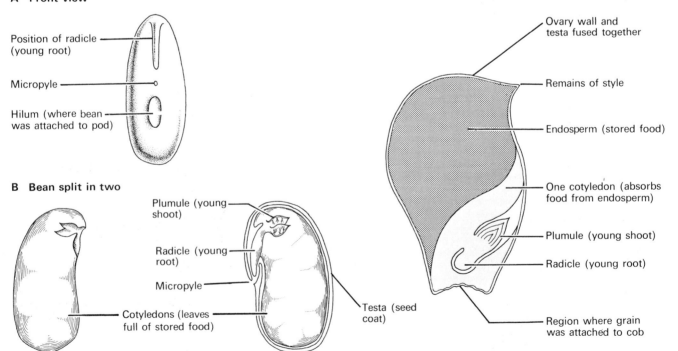

A Front view

Position of radicle (young root)

Micropyle

Hilum (where bean was attached to pod)

B Bean split in two

Plumule (young shoot)

Radicle (young root)

Micropyle

Cotyledons (leaves full of stored food)

Testa (seed coat)

Fig. 68.1 Structure of a kidney bean

Ovary wall and testa fused together

Remains of style

Endosperm (stored food)

One cotyledon (absorbs food from endosperm)

Plumule (young shoot)

Radicle (young root)

Region where grain was attached to cob

Fig. 68.2 Structure of a maize grain (cut in half)

Fig. 68.3 Germination of a kidney bean

Fig. 68.4 Germination of a maize grain

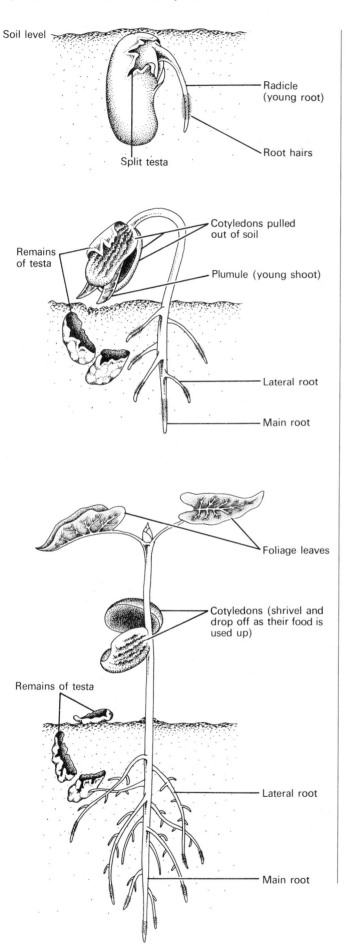

Soil level

Radicle (young root)

Root hairs

Split testa

Remains of testa

Cotyledons pulled out of soil

Plumule (young shoot)

Lateral root

Main root

Foliage leaves

Cotyledons (shrivel and drop off as their food is used up)

Remains of testa

Lateral root

Main root

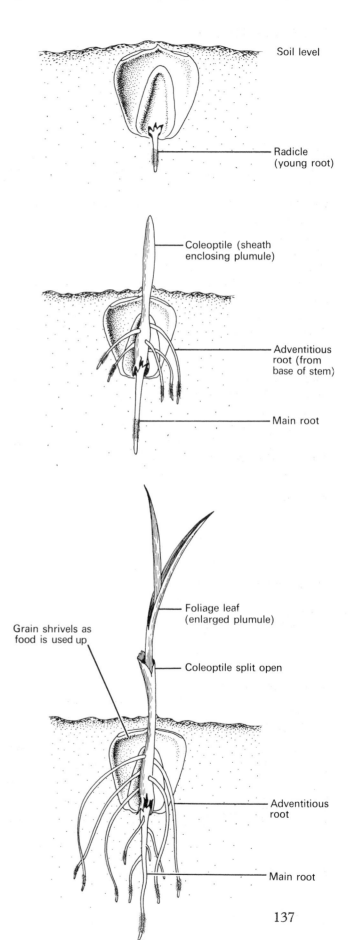

Soil level

Radicle (young root)

Coleoptile (sheath enclosing plumule)

Adventitious root (from base of stem)

Main root

Foliage leaf (enlarged plumule)

Grain shrivels as food is used up

Coleoptile split open

Adventitious root

Main root

137

69

Vegetative reproduction

Many flowering plants are capable of asexual reproduction. The parent plant produces outgrowths which become detached and form new daughter plants. This process is called **vegetative reproduction**.

In blackberries, for example, a stem may curve downwards until it touches the soil. It then forms roots (Fig. 69.1). The stem is now called a **stolon**. The bud at the tip of the stolon, supported by the new roots, forms a daughter plant. When this plant no longer requires food from the parent, the stolon connecting it with the parent dies away.

Strawberries and creeping buttercups (Fig. 69.2) reproduce vegetatively by producing **runners**. Runners grow out over the soil, forming several new plants. When the daughter plants are large enough to be independent, the runner connecting them with the parent dies away.

A **bulb** is a large underground bud whose leaves are swollen with stored water and food. A daffodil (Fig. 69.3) is an example. In spring a leafless flowering stem surrounded by green leaves grows from the centre of the bulb. This growth uses up food stored in the rest of the bulb. After the flower dies the green leaves live on, making food which moves down to the base of each leaf. These swell and form new bulbs. Daughter bulbs develop from buds which form near the outside of the parent bulb. When these reach a certain size they break off the parent bulb and become independent plants.

A **corm** is a very short stem, swollen with stored food, which is situated permanently underground. A gladiolus (Fig. 69.4) is an example. In spring the bud on top of a corm grows a flower and leaves, using up all the food stored in the corm. Later the leaves make food which passes down to form a new corm on top of the shrivelled remains of the old one. Special contractile roots pull the new corm down into the soil. Daughter corms develop from buds on the side of the parent corm. These new corms eventually break off the parent corm and become independent plants.

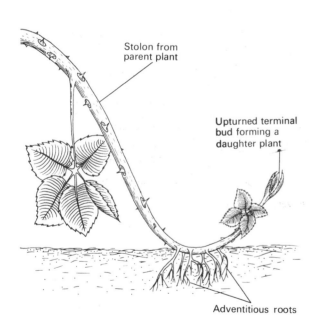

Fig. 69.1 Blackberry stolon

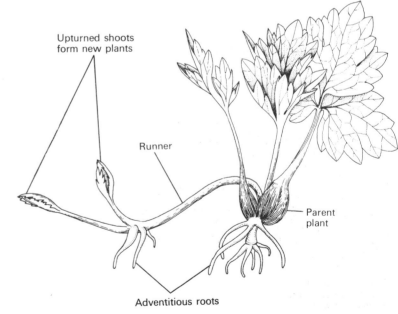

Fig. 69.2 Buttercup runner

A Daffodil bulb

Remains of flower stalk (inside the bulb)

Daughter bulb formed from an outer axillary bud

Brown scale leaves (bases of foliage leaves from two years ago)

Adventitious roots

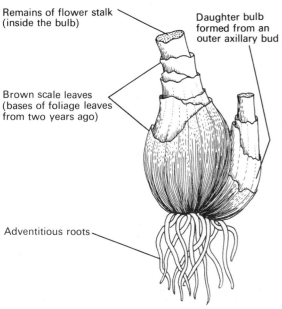

Fig. 69.3 Daffodil bulb

B Diagram of daffodil bulb

Scale leaves

Flower bud

Fleshy scale leaves full of food and water (had green tops one year ago)

Young foliage leaves not attached to flower stalk (will form this year's foliage and next year's food store)

Axillary bud (may form new bulb on side of old one)

Short dome-shaped stem

Adventitious root

A Gladiolus corm as it appears in autumn

Flower stem

Dry scaly bases of this year's leaves

Scars left by this year's leaves

Corm formed by this year's growth

Shrivelling remains of last year's growth

Remnants of corm formed two years ago

Adventitious roots

Contractile adventitious roots (pull new corm down into the soil)

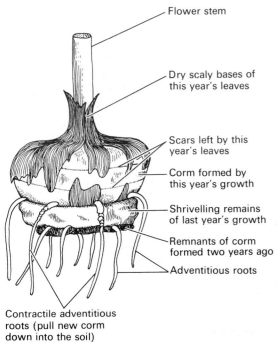
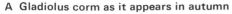

Fig. 69.4 Gladiolus corm

B Diagram of corm ready for planting (vertical section)

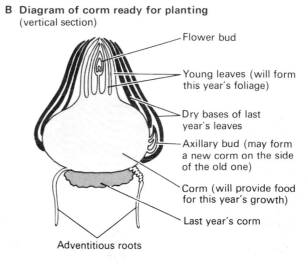

Flower bud

Young leaves (will form this year's foliage)

Dry bases of last year's leaves

Axillary bud (may form a new corm on the side of the old one)

Corm (will provide food for this year's growth)

Last year's corm

Adventitious roots

139

70

Types of variation

People, and all other living things, vary in many ways. In fact no two individuals are ever exactly alike. Humans, for example, all have the same general shape and the same set of body organs. But features such as height, weight, eye and hair colour, the shape of the nose, language, knowledge and skills, body scars, etc., differ from one person to the next. These are examples of individual **variation**.

Children can inherit some of these characteristics from their parents, but not all of them. A child may have the eye or hair colour of one or other of its parents, but there is no chance whatsoever that it will inherit a knowledge of French from its mother, or a scar which its father received in an accident. Characteristics which can be inherited from parents are called **hereditary characteristics**. They include colouring of hair and eyes, shape of the nose, ears, and mouth, and all the other physical features which develop in a child as it grows from a fertilized egg. Characteristics such as language, scars, skills, etc., are called **acquired characteristics**, because they are acquired after birth.

Most acquired characteristics can be changed, but usually hereditary characteristics cannot be altered. It is possible to bleach brown hair, but brown eyes cannot be made blue, and one blood group cannot be changed into another.

Some characteristics, such as height and weight, are described as showing **continuous variation**. This means there are many intermediate forms of them between one extreme and another. People, for example, occur in so many different sizes that it is possible to arrange even a small group into a continuous line from the smallest person to the tallest (Fig. 70.1).

Other characteristics have no intermediate forms or very few of them. These are described as showing **discontinuous variation**. With rare exceptions people are either distinctly male or distinctly female and not somewhere between the two. In most cases people's ears clearly have lobes or do not have them. Some people can roll their tongues into a U-shape, others cannot (Fig. 70.2); there are no stages in between the two extremes.

Fig. 70.1 An example of continuous variation
People can be any size between the two extremes of very short or very tall. In other words height is an example of continuous variation

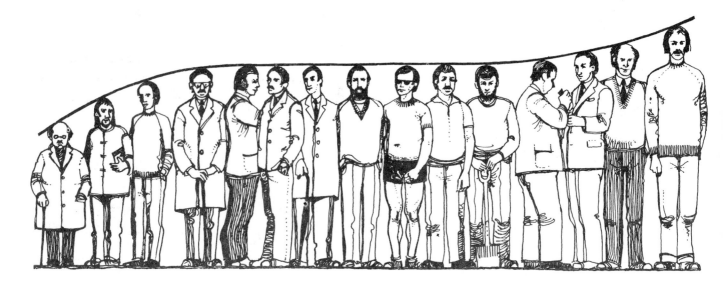

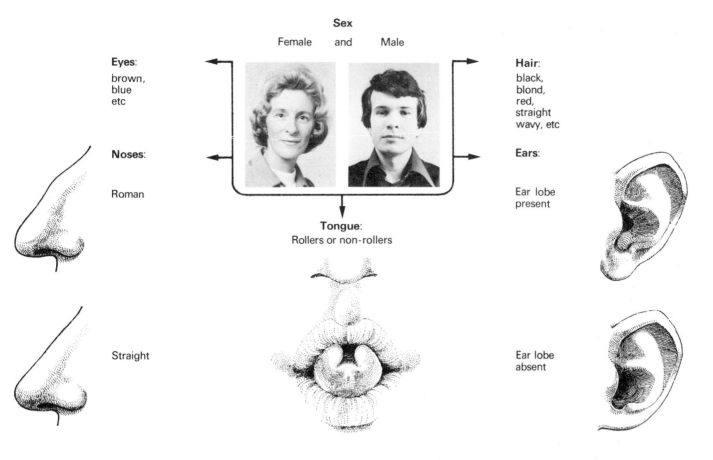

Fig. 70.2   Examples of discontinuous variation in humans

Variation in pea plants

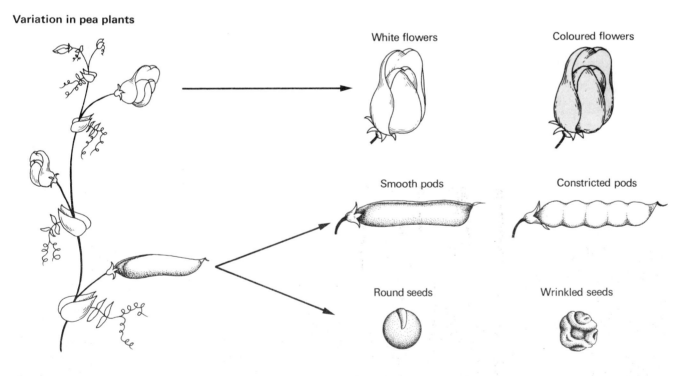

Fig. 70.3   Examples of discontinuous variation in pea plants

71

Mendel's experiments

In 1865 a monk called Gregor Mendel, living in what is now Czechoslovakia, published the first accurate description of the way in which hereditary characteristics pass from parents to their young. Mendel was successful where others failed because of his careful experimental methods.

First, Mendel chose to work with garden peas, because they grow quickly, and their flowers are large enough either to be cross-pollinated, or self-pollinated by hand.

Second, Mendel studied only one or two characteristics at a time. He did this by cross-pollinating plants which differed in some clearly visible way. For example, he studied the inheritance of stem length by cross-pollinating tall plants with dwarf plants. Other characteristics studied in this way are shown in Fig. 71.1.

Third, Mendel always began his experiments with **pure lines**. These are plants which, when self-pollinated, always produce seeds which grow into plants with the same characteristic. For example, pure line tall plants produce *only* tall plants.

Fourth, he studied only those characteristics which show discontinuous variation; that is, when bred together their seeds produce plants of two distinct types, with no intermediate forms. When, for example, pure line tall and pure line dwarf plants are bred together they produce either tall or dwarf plants, with no intermediate sizes.

When two organisms which differ in some way are bred together their young are called **hybrids**. Mendel produced hybrids by cross-pollinating plants. When the parent organisms differ in only *one* way (such as in the length of their stems) the experiment is called a **monohybrid cross**.

To obtain a monohybrid cross between tall and dwarf plants Mendel first had to prevent self-pollination by removing the anthers from a number of tall and dwarf plants before the pollen was released. When the carpels of the plants were ripe he pollinated them with pollen from the *opposite* variety: he dusted stigmas of dwarf plants with pollen from tall plants; and dusted stigmas of tall plants with pollen from dwarf plants (Fig. 71.3). Seeds from these plants produced a new crop of plants which he called the **first filial generation**, or F_1 for short.

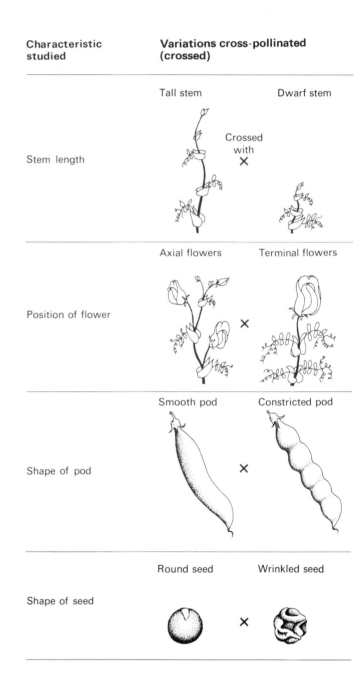

Characteristic studied	Variations cross-pollinated (crossed)
Stem length	Tall stem — Crossed with × — Dwarf stem
Position of flower	Axial flowers × Terminal flowers
Shape of pod	Smooth pod × Constricted pod
Shape of seed	Round seed × Wrinkled seed

Fig. 71.1 Some hereditary characteristics of pea plants studied by Mendel

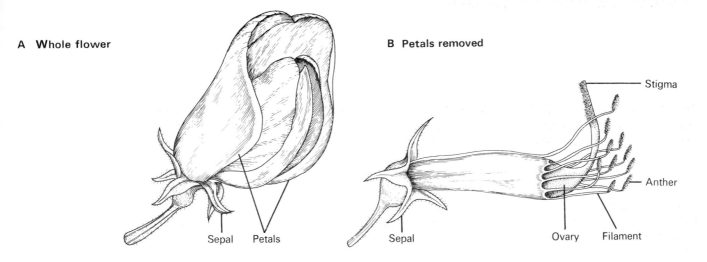

A Whole flower

Sepal   Petals

B Petals removed

Stigma

Anther

Sepal   Ovary   Filament

Fig. 71.2 Structure of a pea flower

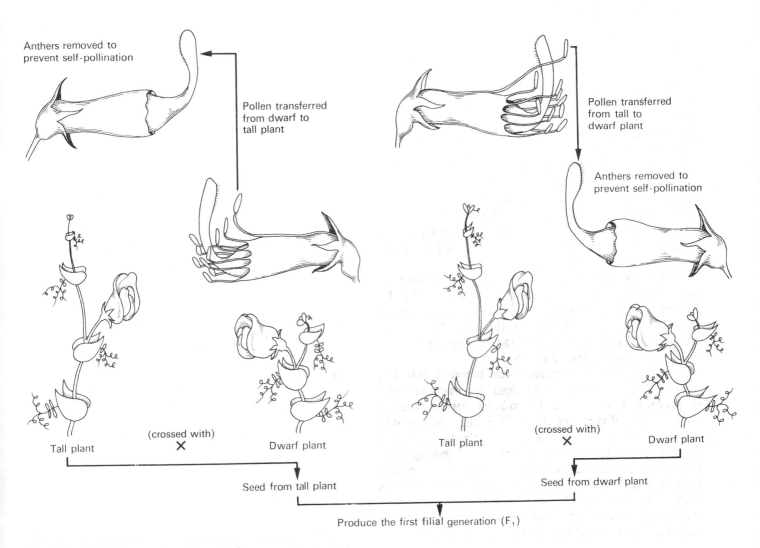

Anthers removed to
prevent self-pollination

Pollen transferred
from dwarf to
tall plant

Pollen transferred
from tall to
dwarf plant

Anthers removed to
prevent self-pollination

Tall plant   (crossed with)
×   Dwarf plant

Tall plant   (crossed with)
×   Dwarf plant

Seed from tall plant

Seed from dwarf plant

Produce the first filial generation (F₁)

Fig. 71.3 Mendel's experimental method

72

Mendel's first law: the law of segregation

Mendel carried out monohybrid crosses to help him find the natural laws which govern the ways in which hereditary characteristics pass from parents to their young.

After cross-pollinating tall and dwarf plants he planted their seeds to obtain the F_1 generation. Without exception all the F_1 plants were tall, whether or not the seeds which produced them came from tall or short parents. Mendel said that, since tallness had 'dominated' dwarfness, tallness should be called the **dominant characteristic.**

Next, Mendel self-pollinated the F_1 plants, collected their seeds, and planted them to obtain the F_2 generation. He found that roughly three-quarters of the F_2 plants were tall, and one-quarter were dwarf, giving a ratio of about 3:1. Since dwarfness had reappeared after 'receding' in the F_1 plants, he called dwarfness a **recessive characteristic** (Fig. 72.1). Many similar experiments produced the same results: the dominant and recessive characteristics appeared in F_2 plants in a ratio of 3:1.

To explain these results Mendel suggested that the bodies of organisms must contain microscopic particles which control hereditary characteristics. He called these particles **factors.** Tallness, for instance, is controlled by a dominant factor. This factor could be represented in a diagram by the capital letter 'T'. Dwarfness is controlled by a recessive factor, which could be represented by a small 't'. Mendel argued that factors must work together in pairs. A pure line tall plant for example, must have the factors TT, and a pure line dwarf must have the factors tt. An F_1 hybrid has the factors Tt. The 'T' factor dominates the 't' factor, making the plant tall.

These ideas were the basis of Mendel's first law, which states that in a cross between plants with contrasting characteristics (such as tallness and dwarfness), the factors segregate (separate) in the F_2 generation. Figure 72.2 explains what this means. When plants reproduce, their pollen and ova receive only *one* factor from each pair of factors. Therefore the pollen and ova of a TT plant get only one T factor each. At fertilization the factors come together again but in a different order, and in the F_2 generation the TT and tt factors of the parents have segregated in a way which produces dominant and recessive characteristics in a ratio of 3:1.

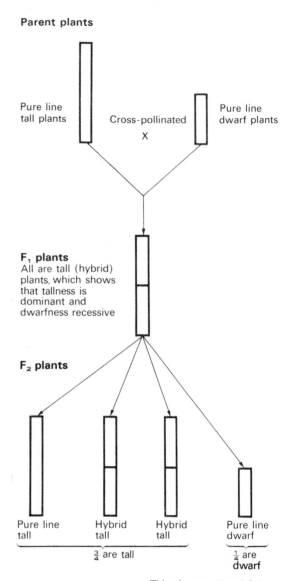

Parent plants

Pure line tall plants

Cross-pollinated X

Pure line dwarf plants

F_1 plants
All are tall (hybrid) plants, which shows that tallness is dominant and dwarfness recessive

F_2 plants

Pure line tall

Hybrid tall

Hybrid tall

Pure line dwarf

$\frac{3}{4}$ are tall

$\frac{1}{4}$ are dwarf

This gives a ratio of 3:1

Fig. 72.1 Diagram of a cross between pure line tall and pure line dwarf plants

Parent plants

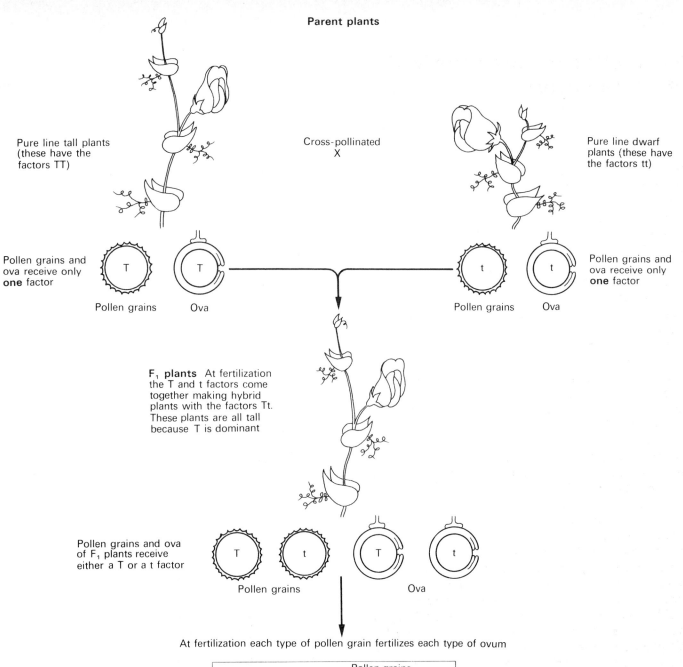

Pure line tall plants (these have the factors TT)

Cross-pollinated X

Pure line dwarf plants (these have the factors tt)

Pollen grains and ova receive only **one** factor

Pollen grains   Ova

Pollen grains   Ova

Pollen grains and ova receive only **one** factor

F₁ plants At fertilization the T and t factors come together making hybrid plants with the factors Tt. These plants are all tall because T is dominant

Pollen grains and ova of F₁ plants receive either a T or a t factor

Pollen grains   Ova

At fertilization each type of pollen grain fertilizes each type of ovum

		Pollen grains	
		T	t
Ova	T	TT Pure line tall plants	Tt Hybrid tall plants
	t	tT Hybrid tall plants	tt Pure line dwarf plants

The ratio of tall to dwarf plants is 3:1

$\frac{1}{4}$ TT
$\frac{1}{4}$ Tt   } $\frac{3}{4}$ Tall plants
$\frac{1}{4}$ tT

$\frac{1}{4}$ tt   } $\frac{1}{4}$ Dwarf plants

Fig. 72.2   Diagram showing how hereditary factors segregate during a monohybrid cross (Mendel's first law)

73

Chromosomes, genes, and genetics

Since Mendel's death, the microscope has revealed that living things are made up of cells, and that the nucleus of a cell contains objects called **chromosomes**. It is now known that the 'factors' which Mendel suggested control hereditary characteristics actually form part of an organism's chromosomes. If chromosomes are observed under a microscope during sexual reproduction, they are seen to behave in the same way as the factors described by Mendel.

First, Mendel said that factors work together in pairs. Just before the production of sex cells (pollen, sperms, and eggs), chromosomes form pairs. Second, Mendel said that factors separate and one from each pair goes into each sex cell. Chromosomes also separate: one from each pair enters each sex cell, so that a sex cell contains half the number of chromosomes in a normal cell. Third, like Mendel's factors, chromosomes form pairs again at fertilization and the normal number is restored. Nowadays, hereditary factors are called **genes**, and the scientific study of heredity is **genetics**.

Each gene controls the development of one or more hereditary characteristics. In each cell of the human body there are 46 chromosomes and between them they contain about 10 000 genes. Some genes control development of body organs; others control visible features such as hair and eye colour, the shape of the nose, mouth, ears, etc.

One of the most popular organisms used in modern genetical experiments is the tiny fruit fly *Drosophila melanogaster*. This fly is chosen for experiments because it breeds quickly (it has a 10-day life cycle) and occurs in many different varieties: some have red eyes, others have white eyes; some have crumpled or vestigial wings (abnormally short wings).

Figure 73.1 is a diagram of a cross between normal-winged and short-winged *Drosophila*. The results follow the same pattern as those achieved by Mendel with pea plants. Figure 73.2 illustrates what happens during this cross to the pair of chromosomes that contain the genes concerned with wing development. It shows how the genes segregate in the F_2 generation according to Mendel's first law.

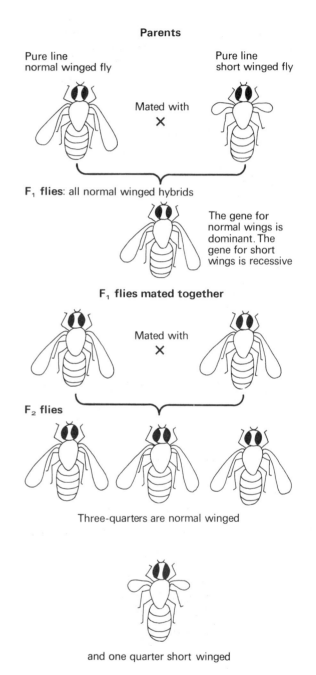

Parents

Pure line normal winged fly

Mated with ✕

Pure line short winged fly

F_1 **flies**: all normal winged hybrids

The gene for normal wings is dominant. The gene for short wings is recessive

F_1 **flies mated together**

Mated with ✕

F_2 **flies**

Three-quarters are normal winged

and one quarter short winged

Fig. 73.1 Diagram showing a cross between short winged and normal winged fruit flies (differences between males and females not shown)

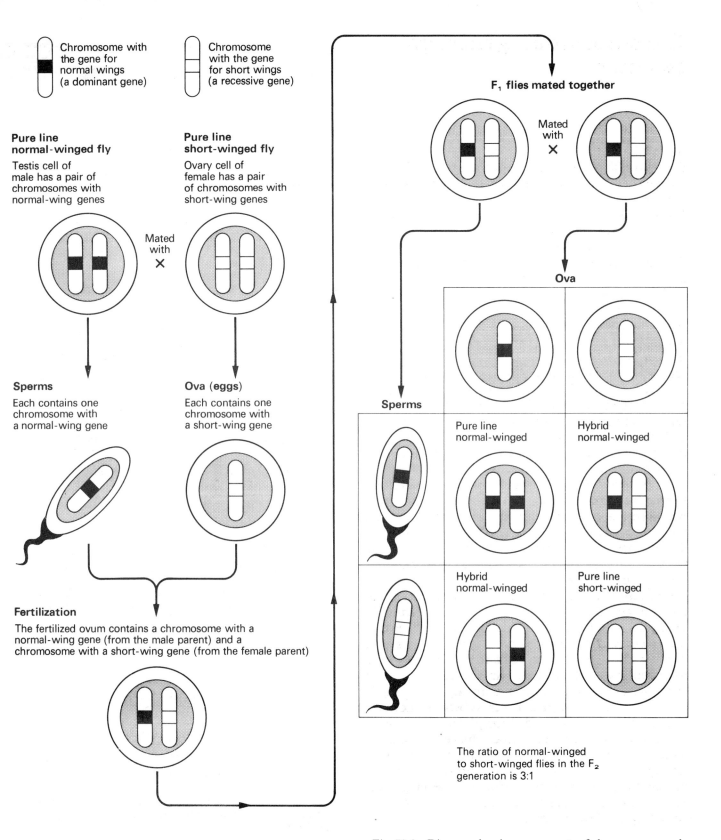

Chromosome with the gene for normal wings (a dominant gene)

Chromosome with the gene for short wings (a recessive gene)

F₁ flies mated together

Mated with ✕

Pure line normal-winged fly

Testis cell of male has a pair of chromosomes with normal-wing genes

Mated with ✕

Pure line short-winged fly

Ovary cell of female has a pair of chromosomes with short-wing genes

Ova

Sperms

Each contains one chromosome with a normal-wing gene

Ova (eggs)

Each contains one chromosome with a short-wing gene

Sperms

Pure line normal-winged

Hybrid normal-winged

Hybrid normal-winged

Pure line short-winged

Fertilization

The fertilized ovum contains a chromosome with a normal-wing gene (from the male parent) and a chromosome with a short-wing gene (from the female parent)

The ratio of normal-winged to short-winged flies in the F₂ generation is 3:1

Fig. 73.2 Diagram showing movements of chromosomes and genes during the cross illustrated in Figure 73.1 (only the pair of chromosomes containing genes which control wing development are shown)

74

Evolution and the record of the rocks

There are more than one and a half million different kinds of living things. Where did they all come from? There is now an enormous amount of evidence that millions of years ago only a few simple organisms existed. From these early forms the huge variety of organisms alive today slowly developed, or **evolved**.

Evolutionary changes occur during reproduction. Organisms produced by sexual reproduction are always slightly different from their parents and, when they are old enough, they in turn produce young slightly different from themselves. Over millions of years the slight differences between one generation and the next eventually produce new types, or species, of organisms.

The most important evidence for evolutionary change comes from a study of **fossils**. Normally when an organism dies its body decays completely. But if by chance a dead body sinks into sediment (mud and sand) at the bottom of a river its soft parts decay, but its hard parts, such as the skeleton, may slowly absorb minerals from the water until they change into stone. A fossil has then been formed. Additional layers of sediment containing more fossilizing bodies gather on the river bed. These press on the deeper layers, hardening them into **sedimentary rocks**. Millions of years later these rocks may be pushed upwards by changes in the earth's crust. The fossils in them are then exposed as the rocks crack open, or as water cuts valleys through them.

The lowest (deepest) layers of sedimentary rock were formed first. They therefore contain the oldest fossils, and the highest (uppermost) rocks contain the most recent fossils. This range of fossils, from the oldest to the most recent, is called the **record of the rocks**. It is a record of the evolutionary changes which have taken place from the earliest forms of life up to the comparatively recent past.

Sometimes a large number of fossils are found which, when arranged in order of age, give an almost complete record of evolutionary change. Figures 74.1 and 74.2 illustrate a small part of the fossil records of the horse and man. If many such fossil records are put together it is possible to construct an **evolutionary tree**.

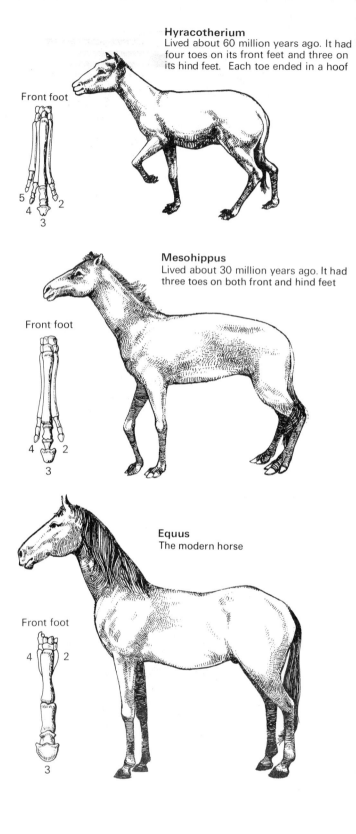

Hyracotherium
Lived about 60 million years ago. It had four toes on its front feet and three on its hind feet. Each toe ended in a hoof

Front foot

5 4 3 2

Mesohippus
Lived about 30 million years ago. It had three toes on both front and hind feet

Front foot

4 3 2

Equus
The modern horse

Front foot

4 3 2

Fig. 74.1 Evolution of the horse. Note the gradual loss of toes and the development of the third toe, which forms the hoof of a modern horse

148

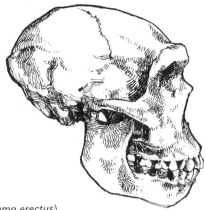

Java ape-man (*Homo erectus*)

Lived between 1 million and 500 000 years ago. Note the receding forehead and chin, and large eyebrow ridges

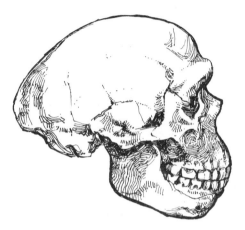

Neanderthal man (*Homo sapiens neanderthalensis*)

Note the higher forehead, elongated skull, smaller eyebrow ridges, and more prominent chin

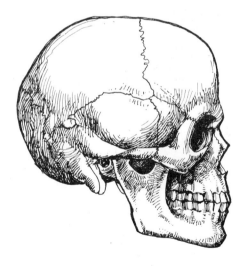

Modern man (*Homo sapiens*)

Note the rounded skull, high forehead, and prominent chin

Fig. 74.2 Evolution of man

75

An evolutionary tree

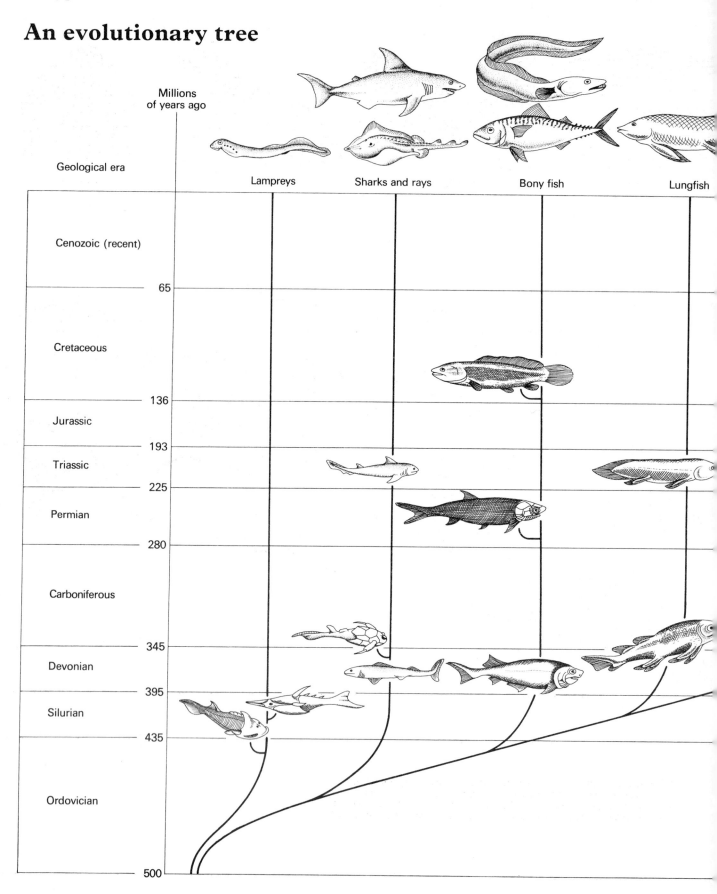

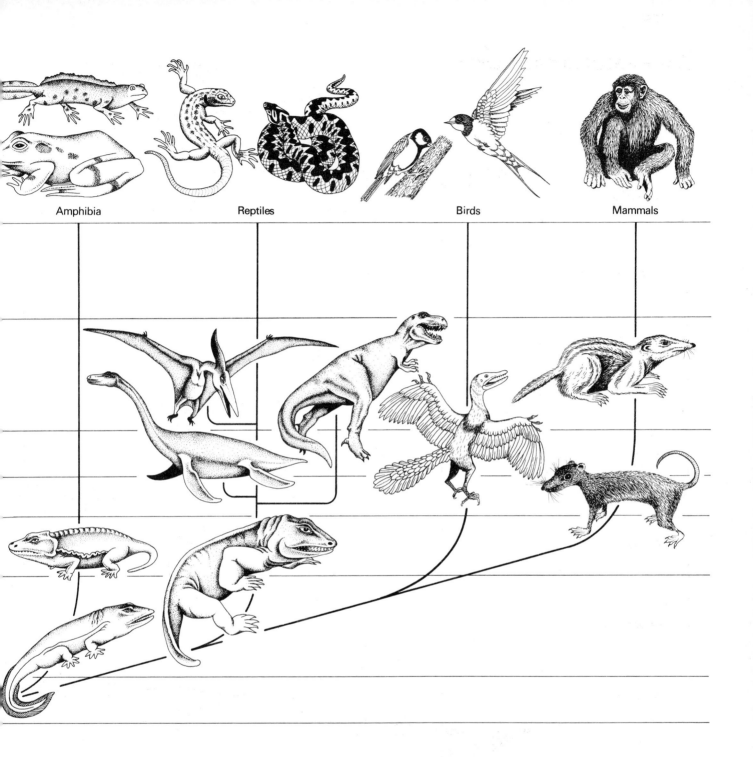

Amphibia   Reptiles   Birds   Mammals

This chart is based on evidence from fossils. It shows that present day
vertebrates evolved from fish which lived more than 400 million years ago.
Some animals have evolved very little. Lungfish, for example, have hardly
changed in 150 million years. All animals below the top line are extinct

76

Evidence of evolution from anatomy and embryology

There is evidence from fossils that mammals and birds evolved from reptiles, that reptiles evolved from amphibia, and that amphibia evolved from fish. **Embryology**, or the study of embryos, provides more evidence that this happened.

Embryos of mammals, birds, reptiles, and amphibia are so similar to fish embryos that it is difficult to tell them apart (Fig. 76.1). All of these, including human embryos, have gill slits, a fish-like heart (with only one atrium and ventricle), fish-like kidneys, and a muscular tail. Human embryos later develop an amphibian heart (with two atria and one ventricle). Then their gill slits disappear and they develop reptilian kidneys. Finally they develop a four-chambered heart, lose their tail, and develop mammalian kidneys. In other words, during embryonic development humans go through some of the stages by which they evolved from fish into amphibia, reptiles, and finally mammals.

Animals such as moles, whales, horses, and birds are entirely different in shape, and live in entirely different ways, but the structure (anatomy) of their bodies has the same basic plan. Their circulatory, nervous, and excretory systems are similar, and their skeletons have roughly the same number and arrangement of bones (Fig. 76.2). The most satisfactory explanation of these similarities is that all these animals evolved from the same ancestor, whose basic anatomy they now share.

In the course of evolution the anatomy of the forelimbs which these animals inherited from their ancestor has become adapted in many different ways: enabling moles to dig, birds to fly, whales to swim, horses to run, etc. The human hand also follows the ancestral plan, but is capable to performing an enormous range of intricate tasks.

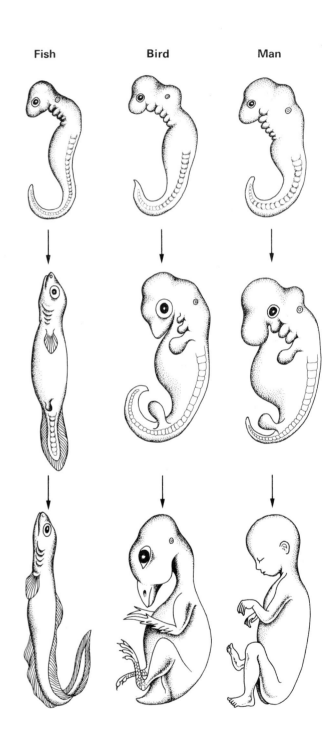

Fig. 76.1 Stages in the development of a fish, a bird, and man. The early stages (top row) are almost identical. Differences appear during later development

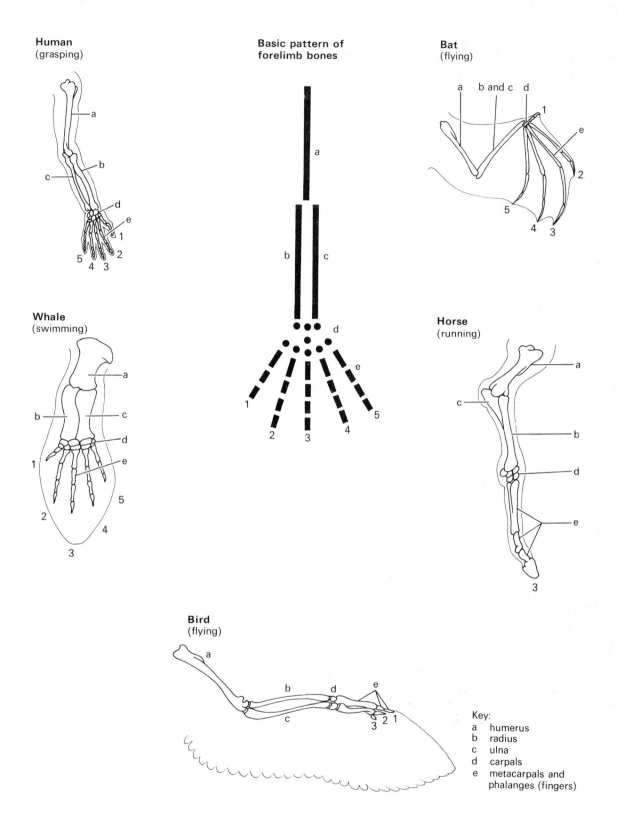

Human
(grasping)

Basic pattern of forelimb bones

Bat
(flying)

Whale
(swimming)

Horse
(running)

Bird
(flying)

Key:
a humerus
b radius
c ulna
d carpals
e metacarpals and
   phalanges (fingers)

Fig. 76.2 The limbs around the outside of the diagram have
different shapes and functions, but share the same basic
structure shown in the central diagram. This is evidence that
all these animals evolved from the same ancestor, whose bone
structure they have inherited

77

The theory of natural selection

The theory of evolution is one way of explaining the presence in the world of countless different kinds of living things. The **theory of natural selection**, put forward by Charles Darwin in 1859, is the most likely explanation of *how* evolution could have occurred; in other words, how one species could have evolved from another. This theory can be summarized as follows:

1. Variation is a characteristic of living things. All living things, even those of the same species, differ from each other in many ways.

2. Living things reproduce at a rate which could at least double their numbers at each generation. Nevertheless, the actual numbers of wild animals and plants remain fairly constant over long periods.

3. Therefore many organisms must fail to survive. There are many reasons for this: predators, lack of food, disease, cold, etc. This means there is a **struggle for survival**, in which many organisms die before they can reproduce.

4. Some variations help organisms to win the struggle for survival. In technical terms, these variations have **survival value**. Organisms with these variations will grow and reproduce while their less favoured fellows die. This process is called **natural selection**, because 'nature' is 'selecting' those best fitted to survive. In other words, natural selection ensures the '**survival of the fittest**'.

5. When surviving organisms reproduce they pass any variations with survival value to their young, who pass these, plus additional 'successful' variations to their young, and so on. Over millions of years this process will lead to the accumulation within a species of more and more variations with survival value. The species will slowly adapt to the conditions in which it lives, and may evolve into a new species.

Darwin showed that selection can change a species by pointing out that man uses **artificial selection** to change animals and plants for his own use. Man produces new varieties by deliberately selecting those which possess the characteristics he desires for breeding. Starting with a single wild variety, man has produced hundreds of different pigeons, dogs, horses, garden flowers, crop plants, etc. If artificial selection can change a species in a few years, it is reasonable to assume that natural selection could have had the same effect over hundreds of millions of years.

Cairn terrier

Chiahuahua

Irish wolfhound

Great Dane

Artificial selection in dogs (*above*) and pigeons (*right*)

Common pigeon

Young fantails

Holle cropper

Brunner cropper

Chinese owl

155

78

Examples of natural selection

A famous example of natural selection favouring one organism rather than another, so that it becomes commoner than its less favoured fellows, is the peppered moth. There are two types of peppered moth: a pale variety which has white wings with black spots, and a dark variety with black wings. The dark variety is the commoner of the two in industrial areas because it is almost invisible to insect-eating birds in smoke-blackened surroundings. The pale variety, however, is clearly visible to predators.

In unpolluted country areas pale wings have survival value since they are almost invisible against a mottled background, such as a tree trunk (see photographs below). Pale moths are therefore commoner than dark moths in country districts.

Charles Darwin discovered an example of natural selection which appears to have produced several new species. In the Galapagos Islands near Ecuador he discovered several previously unknown types of finch-like birds with differently shaped beaks and different feeding habits. Darwin suggested how these birds might have evolved.

Long ago finches from the mainland, where only seed-eating species are found, could have been blown to the Galapagos Islands by strong winds. As their numbers increased they would have to compete for seeds, and some birds would starve. However, since variation is a natural characteristic of all living things, some birds would have slightly different beak shapes. A few of these variations would enable the birds to live on different foods: insects, leaves, fruit, etc. These birds would live to produce young with the same kind of beak, which may eventually have become new species (Fig. 78.1).

Dark and light varieties of peppered moth on a lichen-covered tree in unpolluted woodland. The light variety merges with its background, but the dark moth is easily seen by insect-eating birds

Dark and light varieties of peppered moth on the smoke-stained bark of a tree near an industrial area. The dark variety is camouflaged, but the light one is easily seen by predators

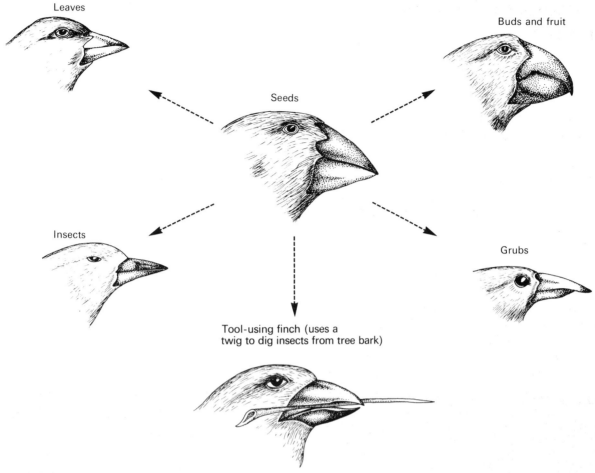

Fig. 78.1 Six of the sixteen types of Galapagos finch. The seed-eating type (centre) is thought to be closest to the seed-eating ancestors from which all the other types have evolved

Leaves

Buds and fruit

Seeds

Insects

Grubs

Tool-using finch (uses a twig to dig insects from tree bark)

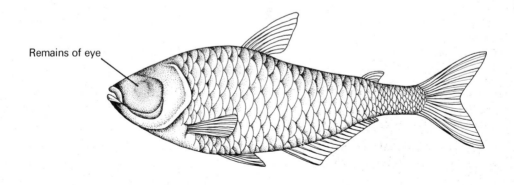

Remains of eye

Fig. 78.2 Blind cave fish. Thousands of years ago fish with eyes were trapped in underground lakes. In time fish without eyes evolved. There are also blind amphibia and insects in caves. These are examples of 'negative evolution', where natural selection has removed useless organs rather than developed new ones

79

Bacteria, disease, and the spread of infection

Bacteria (*singular :* bacterium) are among the smallest and simplest living things. The commonest bacteria are so small that a row of a hundred would just reach across the full stop at the end of this sentence.

Bacteria which cause disease are examples of **germs.** Germs are parasites, and a parasite which harms its host is said to be **pathogenic. Pathology** is the study of the diseases which pathogenic organisms cause.

Bacteria are responsible for some of the deadliest human diseases, such as tuberculosis, leprosy, cholera, diphtheria, pneumonia, and typhoid fever. In addition, bacteria cause ailments which, though less deadly, are nevertheless very unpleasant, such as boils, food poisoning, and bacillary dysentery. Bacteria also cause many plant diseases, some of which damage crops. Pathogenic bacteria harm their hosts by destroying tissues, and by producing poisons called **toxins.**

Under favourable conditions bacteria reproduce by dividing in two every 20 or 30 minutes. Starting with a single bacterium, this rate of division could produce 4000 million, million, million bacteria in one day.

Bacteria, and other germs called **viruses,** are spread from person to person in the following ways:

Droplet infection Germs in the mouth, nose, and lungs can be carried to other people or to water and food in droplets of moisture which leave the body whenever a person breathes out, talks, coughs, or sneezes. Diseases which can be spread in this way include influenza, pneumonia, whooping cough, and diphtheria.

Contaminated food Food and water can be contaminated with germs by sewage, by people with diseases, by insects, birds, and other animals. Typhoid fever, cholera, and food poisoning are spread in this way.

Direct contact Diseases can be spread directly by touching infected people, or objects which they have handled, such as books, coins, door knobs, etc. Diseases spread in this way are called **contagious** diseases. Examples include smallpox, measles, and tuberculosis.

Animals Animals which spread infection are called **vectors.** Houseflies, for example, can carry typhoid, cholera, and dysentery germs on their bodies or in their faeces. Yellow fever and malaria germs are spread by mosquitoes as they suck blood.

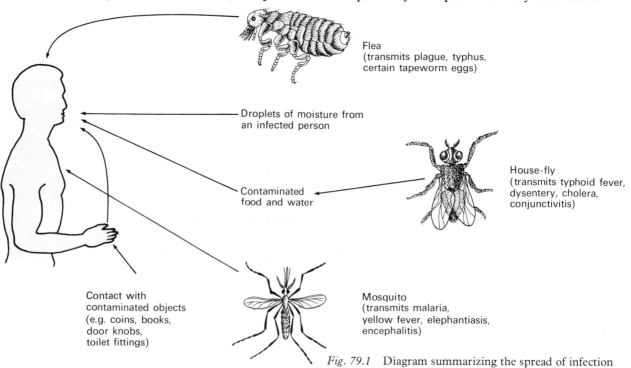

Flea
(transmits plague, typhus,
certain tapeworm eggs)

Droplets of moisture from
an infected person

House-fly
(transmits typhoid fever,
dysentery, cholera,
conjunctivitis)

Contaminated
food and water

Contact with
contaminated objects
(e.g. coins, books,
door knobs,
toilet fittings)

Mosquito
(transmits malaria,
yellow fever, elephantiasis,
encephalitis)

Fig. 79.1 Diagram summarizing the spread of infection

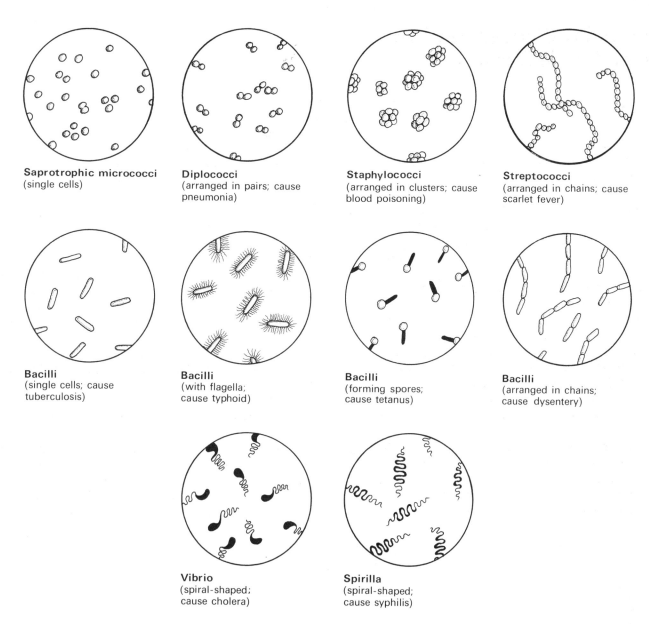

Saprotrophic micrococci
(single cells)

Diplococci
(arranged in pairs; cause pneumonia)

Staphylococci
(arranged in clusters; cause blood poisoning)

Streptococci
(arranged in chains; cause scarlet fever)

Bacilli
(single cells; cause tuberculosis)

Bacilli
(with flagella; cause typhoid)

Bacilli
(forming spores; cause tetanus)

Bacilli
(arranged in chains; cause dysentery)

Vibrio
(spiral-shaped; cause cholera)

Spirilla
(spiral-shaped; cause syphilis)

Fig. 79.2   Some types of bacteria

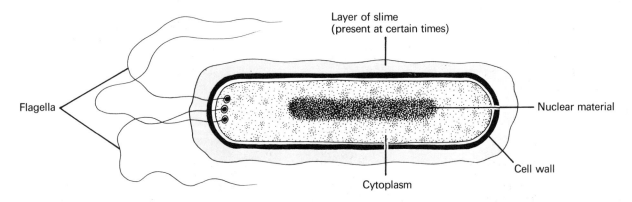

Layer of slime
(present at certain times)

Flagella

Nuclear material

Cell wall

Cytoplasm

Fig. 79.3   Structure of a bacterium

80

The body's defences against infection

The body has many defences against germs and other parasites. Together these defences are called **natural immunity**. Here are some examples:

The respiratory system Dust and germs breathed in through the nose do not usually reach the lungs: they are trapped in sticky mucus which covers the walls of the nasal passages, windpipe, and lungs (Fig. 80.2A). Trapped dirt and germs are then carried by cilia to the oesophagus, where they are swallowed and eventually passed out of the body in the faeces.

The skin The surface of the body is covered with skin, which consists of several layers. The outer layer is made up of dead cells and is called the **cornified layer** (Fig. 80.2B). As fast as these dead cells wear away or are damaged they are replaced from below by a region of live growing cells called the **Malpighian layer**. The dead cornified layer is kept supple, water-repellent, and mildly antiseptic by an oily substance called **sebum**, which is produced by **sebaceous glands** in the hair follicles. The skin therefore acts as a waterproof, germ-proof, self-repairing barrier against germs and dirt.

The eyes are protected from infection by a thin skin called the **conjunctiva**, which is continuously bathed in an antiseptic fluid produced by the **tear glands** at the corners of the eyes. The blink reflex shuts the eyelids to protect the eyes against dust, bright light, flying insects, etc. The eyelashes also provide some protection.

The stomach Many germs are unavoidably swallowed with food and drink. Fortunately these germs are usually harmless, but in any case the majority are killed by by stomach acid and digestive enzymes (Fig. 80.2C).

The examples of natural immunity described so far may be thought of as the body's first line of defence. But if these defences are broken, as happens when the skin is cut, grazed, or burned, or if large numbers of germs are inhaled or swallowed, then the body's second line of defence comes into operation. Second line defences are controlled by the blood.

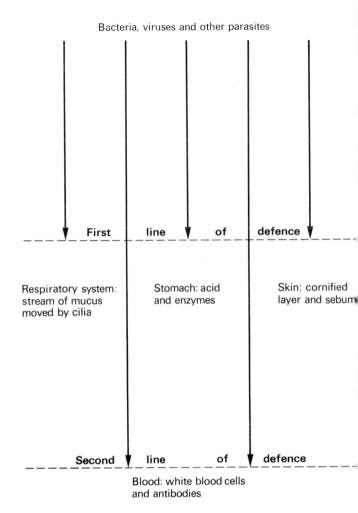

Fig. 80.1 Diagram summarizing natural immunity

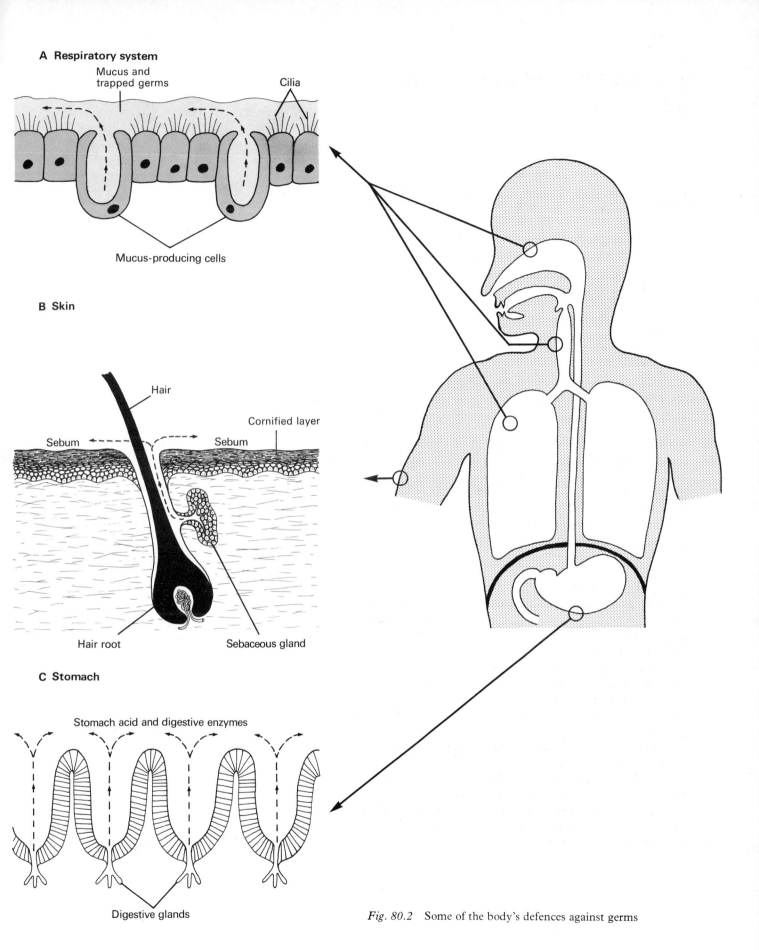

A Respiratory system

Mucus and trapped germs

Cilia

Mucus-producing cells

B Skin

Hair

Cornified layer

Sebum

Sebum

Hair root

Sebaceous gland

C Stomach

Stomach acid and digestive enzymes

Digestive glands

Fig. 80.2 Some of the body's defences against germs

81

More defences against infection

Bleeding and clot formation When the skin is damaged bleeding occurs for a time. This washes germs and dirt out of the wound. Then chemical reactions in the blood cause it to become solid and form a **blood clot**, which plugs the wound. This clot not only stops bleeding: it keeps germs and dirt out of the wound until the skin has healed.

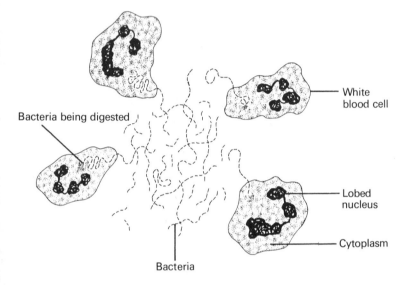

Fig. 81.1 Phagocytic white blood cells destroying bacteria

Action of phagocytes Phagocytes are white blood cells which destroy invading germs by digesting them (Fig. 81.1). Phagocytes are not restricted to the blood-stream. They can leave blood vessels (Fig. 32.2D) and kill germs in wounds and tissue fluid. In addition phagocytes in lymph nodes kill germs which reach the lymph. Germs and the poisons (**toxins**) they produce are destroyed by **antibodies**, which are chemicals produced by the body during an infection.

Antibodies Substances such as germs and toxins which stimulate the body to make antibodies are called **antigens**. There are four types of antibody, and they each help destroy antigens in a different way (Fig. 81.2). **Opsonins** are antibodies which stick to the surface of germs. This makes them more likely to be attacked by phagocytes. It is as if opsonins make germs more 'appetizing'. **Lysins** are antibodies which dissolve germs. **Agglutinins** stick germs together, so that they cannot penetrate cells or reproduce properly. **Antitoxins** make toxins harmless. Antibodies can remain in the blood for years after an infection has disappeared, making the body immune to further infection from the same germs.

Ready-made antibodies It is now possible to make antibodies artificially. These can be injected into the body to help it fight an infection. Diphtheria antibody, for example, is made by injecting horses with diphtheria toxin. The horse produces antitoxins, which are easily extracted from its blood, and can then be injected into humans who are suffering from the disease.

Vaccines Vaccines are liquids containing either dead germs, or harmless live germs which are very similar to a type which causes a serious disease. When a vaccine is injected into the blood, the body makes antibodies as if it were suffering from an infection. The antibodies remain in the blood, making the body immune to a real infection.

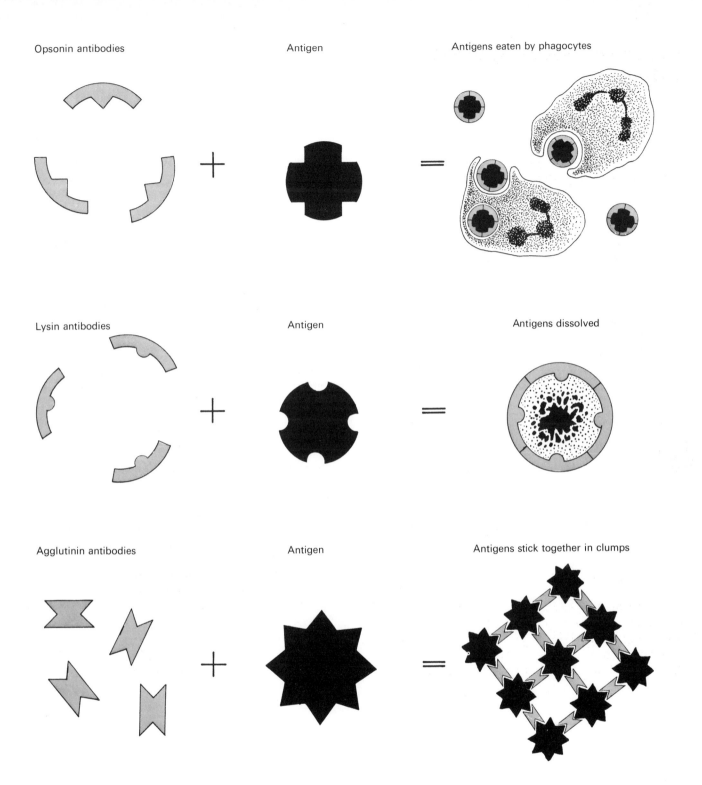

Opsonin antibodies   Antigen   Antigens eaten by phagocytes

Lysin antibodies   Antigen   Antigens dissolved

Agglutinin antibodies   Antigen   Antigens stick together in clumps

Fig. 81.2 Diagram showing how certain antibodies work
Antibody molecules are not really shaped like those in the
diagram. The shapes indicate that each antibody will react with
only one specific antigen

163

82

Smoking and health

There is now an enormous amount of evidence that the following diseases occur more often in smokers than non-smokers: lung cancer, emphysema (thinning and weakening of lung tissue), cancer of the mouth, throat, larynx, gullet, bladder, and pancreas, coronary thrombosis (blockage of arteries to the heart), angina pectoris (pain due to narrowing of arteries to the heart), and chronic bronchitis. In addition smoking appears to delay the healing of stomach ulcers; it reduces the senses of taste and smell; slows down reflexes (which makes smokers more prone to accidents); and gives the breath and clothes an unpleasant smell. Pregnant women who smoke tend to have smaller babies than non-smokers, and their babies are more likely to be born dead, or die a few days after birth.

Evidence that smoking causes lung cancer comes from three main sources. First, world-wide investigations show that people with lung cancer are more often heavy smokers than non-smokers. Second, if people's smoking habits are studied throughout their lives, and the causes of their deaths are recorded, it is found that the more cigarettes they smoked the greater the chances that they died of lung cancer (Fig. 82.2A). Third, the chances of developing lung cancer are reduced if people give up smoking, and continue to decline as the length of time increases since they gave up smoking (Fig. 82.2B).

The method of smoking affects the danger to health. Cigarettes seem to be the most hazardous way of smoking. Pipes are less risky, and cigars seem to be the least dangerous (Fig. 82.2C).

A study of hospital patients has shown that the chances of developing bronchitis and coughing with phlegm are five times greater among smokers than non-smokers (Fig. 82.2D). Furthermore, smokers are twice as likely to die from coronary heart disease as non-smokers (Fig. 82.2E). People who give up smoking greatly reduce their chances of developing these diseases.

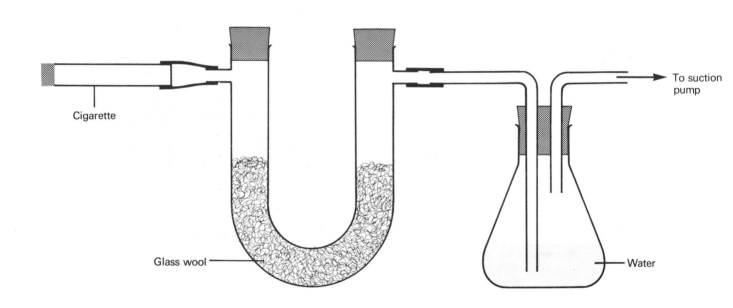

Fig. 82.1 Apparatus for investigating the contents of cigarette smoke
Suck smoke from at least five cigarettes through the apparatus, then examine the appearance and smell of the glass wool and water

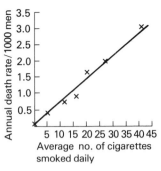

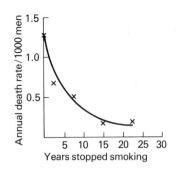

A Death rate from lung cancer among men smoking different numbers of cigarettes each day

B Death rate from lung cancer among men who gave up smoking cigarettes

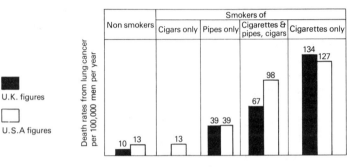

C The effect of smoking tobacco in different ways

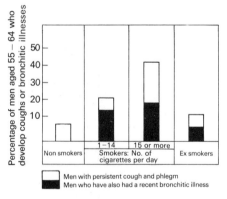

D The effect of smoking on the chance of developing bronchitis or coughing with phlegm

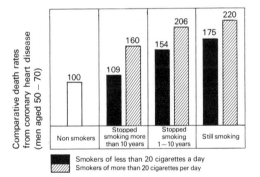

E The relationship between smoking and coronary heart disease

Fig. 82.2 The effects of smoking on health (figures taken from 'Mortality in relation to smoking: ten years' observations by British doctors', by R. Doll and A. B. Hill, *British Medical Journal*, Vol. 1, 1964). The statistics in these graphs refer to men because in the past men have tended to smoke more heavily than women. But women who smoke are liable to contract the same illnesses as men

83

Soil

Shake a few grams of soil in a test-tube of water, allow it to settle, and compare the result with Figure 83.1. The soil will separate into **humus**, which floats at the top of the test-tube, and **mineral particles** arranged in order of size, the largest at the bottom.

Humus is the decayed remains of plants and animals. Plant cellulose and xylem form fibrous humus (like that at the top of the test-tube in Figure 83.1). Animals decay into a liquid which covers mineral particles and sticks them together into clumps called **crumbs** (Fig. 83.2). Humus fibres and soil crumbs help make soil fertile. Air spaces between the crumbs allow water to drain away so the soil does not become waterlogged, and allow air to reach plant roots. Humus retains some moisture and absorbs dissolved minerals essential for healthy plant growth. Humus also binds mineral particles together so they are not blown away in high winds.

The mineral particles in soil are tiny pieces of rock. They are formed from rock shattered by frost and plant roots, and reduced to fragments by rain water made acid by carbon dioxide in the air. This dissolves soluble substances in them. Rock fragments are broken down further as they roll down hillsides and are washed away in rivers (Fig. 83.3). Gradually they are reduced to sand, silt, and clay.

Loam is the most fertile type of soil. It consists of about 50 per cent sand, 30 per cent clay, 20 per cent humus, and has well developed soil crumbs. Loam warms quickly in spring, can be ploughed easily, is well aerated, and drains freely but retains sufficient water and minerals for healthy plant growth.

Sandy soils have little humus, hardly any clay, and have poorly developed soil crumbs. They dry out quickly in a drought, and as rain drains quickly through them it washes away dissolved minerals, and reduces fertility. Sandy soils are improved by adding humus in the form of peat and manure.

Clay soils have little humus or sand. They drain slowly, and when they are wet they contain no air spaces. They dry into hard lumps. Clay soils are improved by deep ploughing, and by adding humus, sand, and lime. Lime joins the fine clay particles into clumps like soil crumbs. This improves drainage.

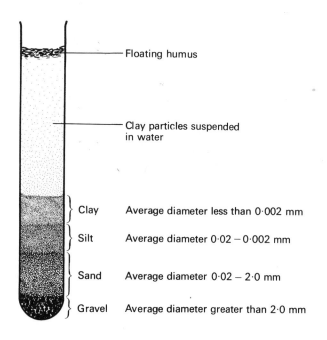

Floating humus

Clay particles suspended in water

Clay — Average diameter less than 0·002 mm

Silt — Average diameter 0·02 — 0·002 mm

Sand — Average diameter 0·02 — 2·0 mm

Gravel — Average diameter greater than 2·0 mm

Fig. 83.1   Results of a soil sedimentation test

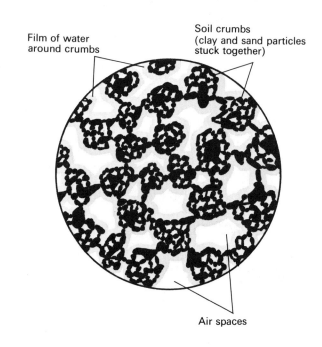

Film of water around crumbs

Soil crumbs (clay and sand particles stuck together)

Air spaces

Fig. 83.2   Magnified view of soil crumbs

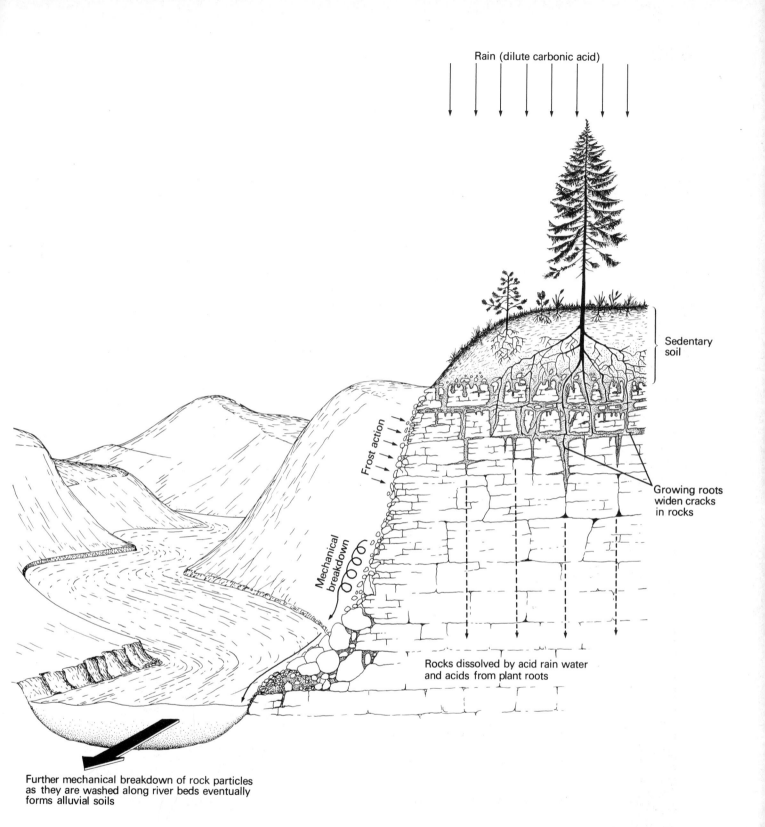

Rain (dilute carbonic acid)

Sedentary
soil

Frost action

Growing roots
widen cracks
in rocks

Mechanical
breakdown

Rocks dissolved by acid rain water
and acids from plant roots

Further mechanical breakdown of rock particles
as they are washed along river beds eventually
forms alluvial soils

Fig. 83.3   Some of the ways in which soil forms

167

84

The carbon cycle

Living things are constantly taking in oxygen, carbon dioxide, water, and minerals. But these materials are never used up completely because as fast as they are absorbed they are replaced by photosynthesis, respiration, excretion, and decay. In other words, there is a **balance of nature**, in which losses equal replacements, and materials are used again and again. In nature materials move in **cycles** (that is they circulate). The **carbon cycle** is an example.

During the day carbon dioxide is absorbed by plants as a raw material of photosynthesis. But plants never run out of carbon dioxide because it is constantly replaced in the following ways.

Respiration Some of the food made by photosynthesis is respired by plants for energy, a process which releases carbon dioxide into the air. Plants are eaten by herbivores, who respire this food and release more carbon dioxide. The same thing happens when herbivores are eaten by carnivores.

Decay After death the bodies of organisms are decomposed and absorbed by bacteria and fungi. These organisms respire some of the material they absorb, which produces carbon dioxide gas.

Combustion (burning) Combustion, or burning, of inflammable materials results in the release of carbon dioxide. Combustion can form part of the carbon cycle in the following ways. Carbon absorbed by a tree during photosynthesis and used to build woody tissue in its trunk will be returned to the atmosphere if the tree is burned as fuel. Over millions of years the bodies of dead organisms have produced fossil fuels such as coal, oil, and natural gas. Some of these organisms were plants which took **carbon** from the air during photosynthesis, and some were animals which fed on the plants, or on other animals. When these fossil fuels are burned today, they release carbon atoms which were trapped by photosynthesis in plants that lived millions of years ago.

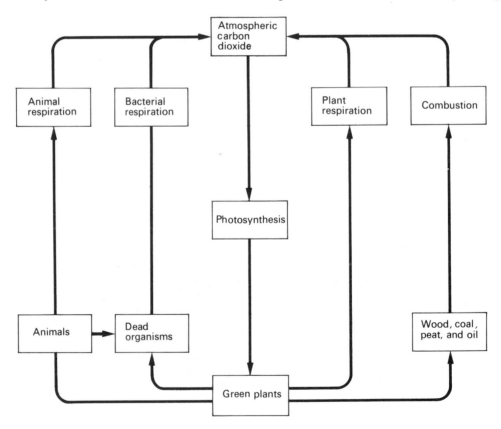

Fig. 84.1 Summary of the carbon cycle

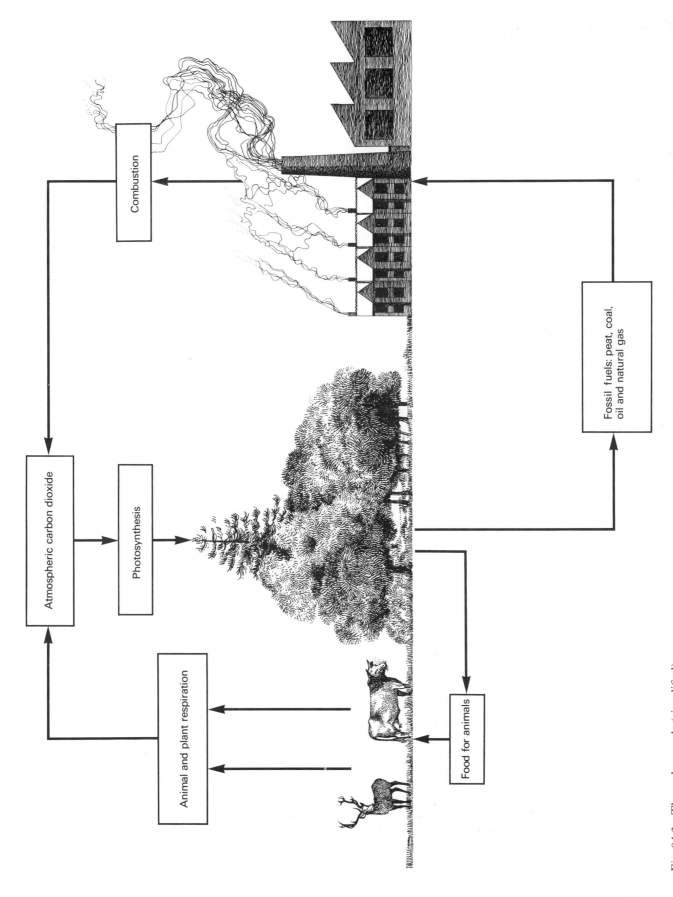

Fig. 84.2 The carbon cycle (simplified)

169

85

The nitrogen cycle

The nitrogen cycle is another example of the continuous circulation of substances through nature. Life cannot exist without nitrogen: it is an essential part of all proteins. Air is four-fifths nitrogen gas, but neither plants nor animals can absorb it in this form. Plants can take up nitrogen only as nitrates (compounds of nitrogen and oxygen) which they absorb from the soil, and animals can only obtain nitrogen by eating plants, or animals which eat plants. Consequently, animals and plants depend on the following processes, which replace soil nitrates as fast as plants absorb them.

Lightning An extremely high temperature is generated during a flash of lightning – high enough to combine nitrogen and oxygen gases into oxides of nitrogen. These dissolve in rain water, soak into the soil, and form nitrates.

Nitrogen-fixing bacteria These are bacteria which use carbohydrate and nitrogen gas to form substances that eventually form soil nitrates. Some of these bacteria obtain carbohydrates from humus. Others live in swellings called **root nodules** on the roots of leguminous plants like peas, beans, and clover. The bacteria obtain carbohydrates from the plants, and release nitrates into the plant tissues and into the soil.

Nitrifying bacteria As bacteria and fungi decompose the dead bodies and faeces of organisms they release ammonia, which is converted into nitrates by nitrifying bacteria. This happens in two stages: **nitrite bacteria** combine ammonia with oxygen to form nitrites, and **nitrate bacteria** combine nitrites with more oxygen to form nitrates.

Denitrifying bacteria Most soils contain denitrifying bacteria. These obtain their energy by breaking down nitrates into nitrogen and oxygen gases, which escape into the air. Fortunately, denitrifying bacteria are only active in waterlogged conditions. Therefore it is good farming practice to improve drainage in soils where flooding is likely.

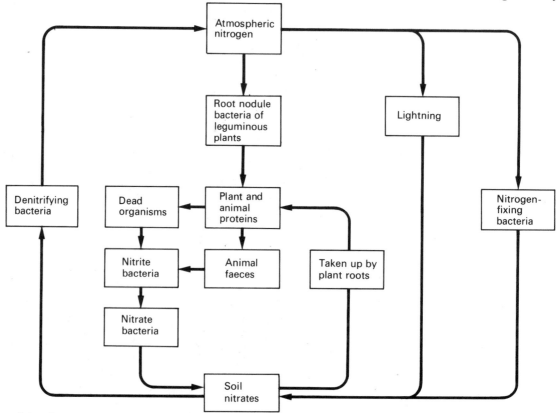

Fig. 85.1 Summary of the nitrogen cycle

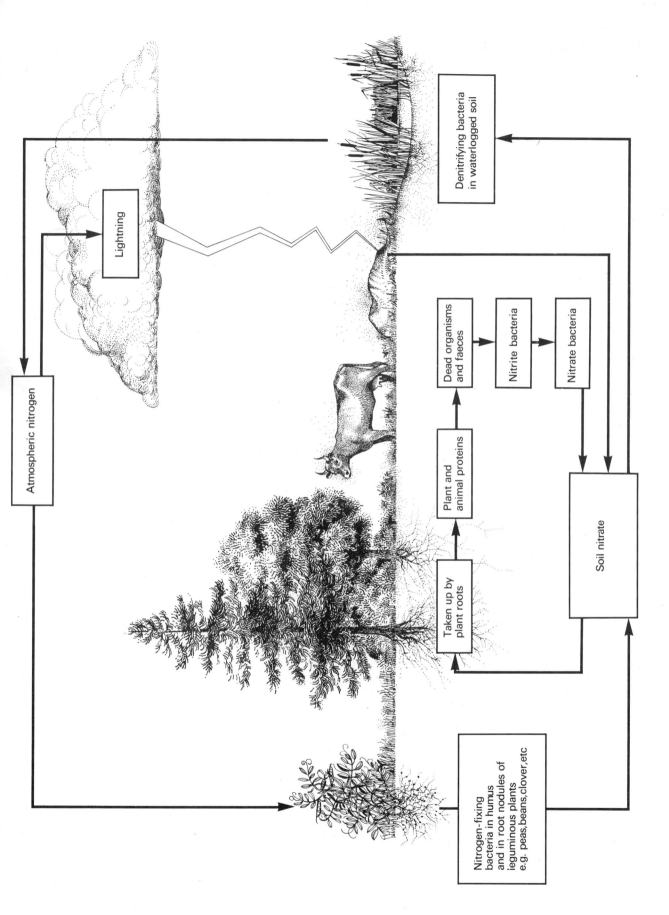

Fig. 85.2 The nitrogen cycle (simplified)

171

86

Food chains and the pyramid of numbers

What would happen if the sun went out? Ignoring for the sake of argument the fact that the world would quickly freeze, lack of sunlight would have the following effects on living things:

1. Green plants would die, since they depend on sunlight to make food by photosynthesis.

2. Herbivores (e.g. rabbits and cattle) would die next, since they eat plants.

3. Carnivores (e.g. foxes and lions) would be the next to die, since they eat other animals.

4. Parasites (e.g. tapeworms) would die with their hosts, since they obtain food from other living things.

5. Scavengers (e.g. carrion crows and maggots) eat dead animals and plants, and decomposers (e.g. bacteria and fungi) live by decomposing dead organisms. These two types of organism would be the last to die.

This means that all living things depend on sunlight: plants depend on it to make food; herbivores depend on it because they eat plants; carnivores depend on it because they eat herbivores; parasites depend on it because they live off other organisms; scavengers and decomposers depend on it because they consume all types of dead organism. Energy, in the form of food, passes from one organism to the next. This leads to the formation of **food chains**.

All food chains begin with green plants. These are called the **producers** of the food chain because they make food by photosynthesis. All other organisms in a food chain are called **consumers**. Herbivores are **primary consumers** because they feed on plants; carnivores which feed on primary consumers are **secondary consumers**. Carnivores which eat secondary consumers are **tertiary consumers**. Parasites, scavengers, and decomposers obtain food from all parts of a food chain (Fig. 86.2).

In all food chains there is a reduction in the number of organisms from the producers at the base of the chain, to the last consumer at the top. This gives rise to a **pyramid of numbers**. Figure 86.1 shows one example of a pyramid of numbers. The single herring at the top of the pyramid eats thousands of small animals called zooplankton every day, and these zooplankton eat millions of even smaller plants called phytoplankton, which form the base of the pyramid.

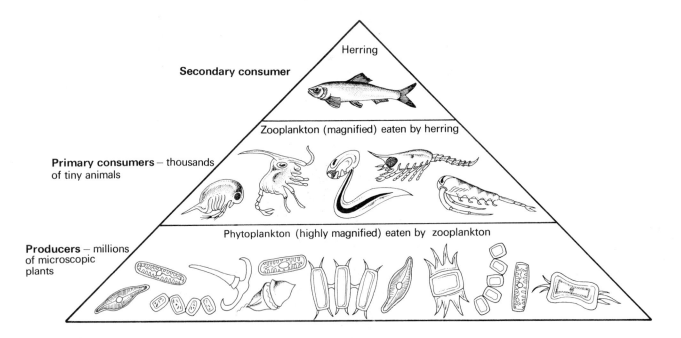

Secondary consumer

Herring

Zooplankton (magnified) eaten by herring

Primary consumers – thousands of tiny animals

Phytoplankton (highly magnified) eaten by zooplankton

Producers – millions of microscopic plants

Fig. 86.1   A pyramid of numbers (or food pyramid)

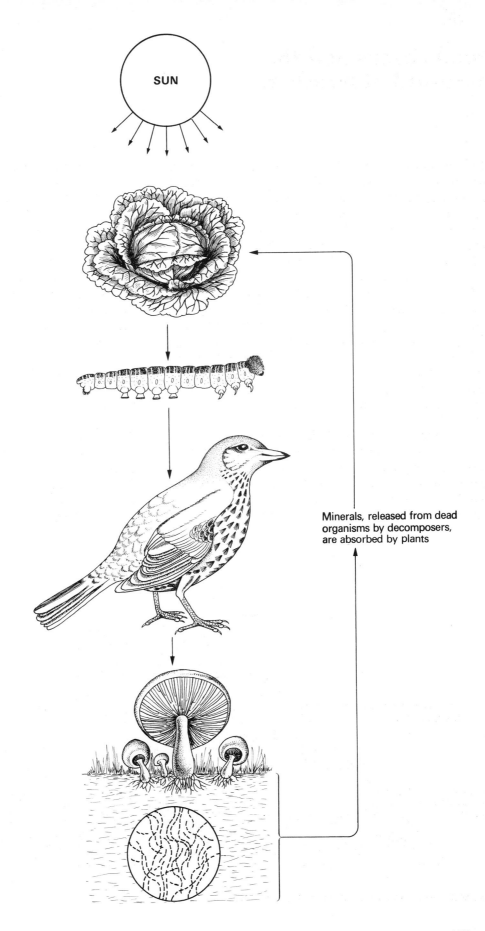

Sunlight
The source of energy
for photosynthesis

SUN

Producers
Green plants (e.g. cabbage)
make food by photosynthesis

Primary consumers
Herbivores (e.g. caterpillar)
eat plants

Secondary consumers
Carnivores (e.g. thrush)
eat herbivores

Minerals, released from dead
organisms by decomposers,
are absorbed by plants

Decomposers
Organisms such as fungi
and certain bacteria
(saprophytes) decompose
dead plants, animals, and
droppings

Fig. 86.2   A simple food chain

87

Food webs, communities, and ecology

Food chains are never as simple as those illustrated on the previous page, because consumers rarely eat only one type of food. Zooplankton, for example (Fig. 86.1), are eaten by prawns, shellfish, and whalebone whales as well as by herring. Food chains therefore interconnect at many points, making what is known as a **food web**. Figure 87.1 shows a simple food web, and Figure 87.2 shows a tiny part of the immensely complex food web that exists in the sea.

The organisms which make up a food web are **interdependent**. This means they depend on each other, especially for food. Any group of interdependent organisms which live together in one place, such as a rock pool, is called a **community**. The place where a community lives is called its **habitat**. A rock pool is the habitat of a community made up of crabs, sea anemones, marine worms, seaweeds, etc. Ponds, moorlands, and hedgerows are other habitats.

The complicated relationships between the members of a community, and between the community and its habitat, are called an **ecosystem**. An ecosystem is made up of all the producers, consumers, scavengers, and decomposers in a community; the rocks, soil, water, and air which make up its habitat; together with materials such as carbon and nitrogen which constantly circulate in the community and its habitat.

The scientific study of ecosystems is called **ecology**. An ecologist studies the organisms in a community to trace the complicated ways in which they depend on one another. But one of the most important jobs of an ecologist is to study man's effect on living things.

Prehistoric men lived by hunting wild animals, gathering fruit, etc. This way of life had little effect on the balance of nature. In the past two hundred years, however, man has changed his surroundings in many ways: by damming rivers, building roads and cities, replacing natural communities with agricultural crops and animals, and, most important of all, by producing poisons and other harmful substances known as **pollutants**.

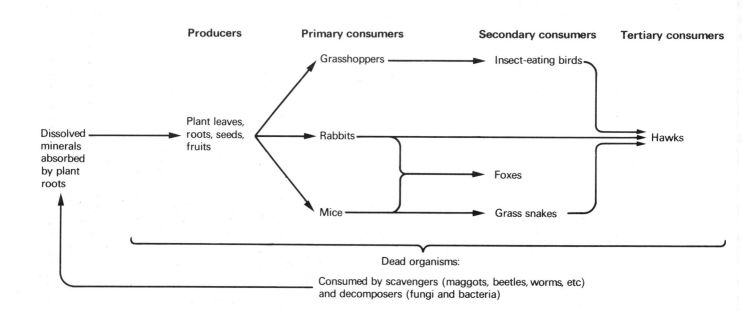

Fig. 87.1 A simple food web

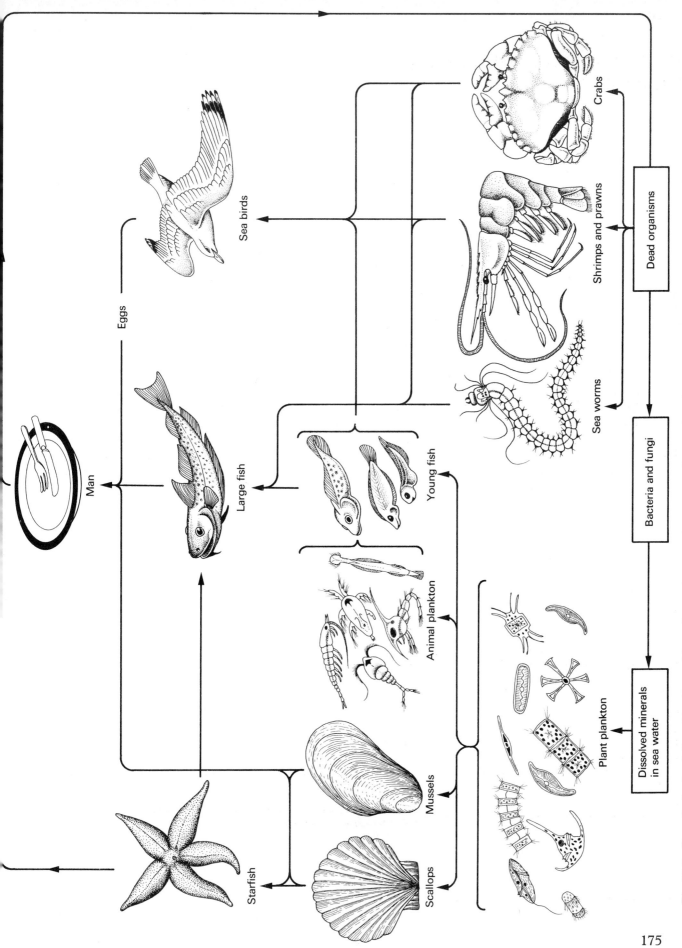

Fig. 87.2   Part of a food web found in the sea. Name the producers; primary, secondary, and tertiary consumers; scavengers; decomposers

Crabs

Shrimps and prawns

Dead organisms

Sea birds

Sea worms

Eggs

Bacteria and fungi

Large fish

Man

Young fish

Animal plankton

Dissolved minerals in sea water

Plant plankton

Mussels

Starfish

Scallops

175

88

Air pollution

A substance becomes a **pollutant** when its presence in air, water, or soil harms living things. The majority of pollutants are harmful for a while but soon become harmless. If sewage, for example, is discharged into the open sea it is quickly made harmless by bacteria, which decomposes it; and poisonous gases from chimneys quickly disperse in the air, and then change into less harmful substances. But a number of pollutants remain poisonous for a very long time. Mercury and cyanide compounds are examples.

The main cause of air pollution is the burning of coal and oil. This happens in factories, in houses, and in the engines of cars, buses, aeroplanes, etc.

The smoke produced by burning contains tiny particles of dust, which are mainly carbon. The dust settles on plant leaves, limiting photosynthesis by cutting out light. It also limits transpiration and gaseous exchange by blocking up stomata. People inhale the dust produced by smoke, and, if it is thick enough, it may aggravate ailments like bronchitis.

The gas called sulphur dioxide is another pollutant produced by burning. Sulphur dioxide gas in the atmosphere comes mainly from coal-burning power stations. There is usually at least one of these in most big cities. As sulphur dioxide passes into the air it reacts with water vapour and changes into sulphuric acid. This damages people's lungs and the leaves of plants.

Petrol and diesel engines, especially when they are not running properly, produce fumes which gather in city streets. These fumes contain oxides of nitrogen and certain lead compounds. Once lead gets into the body it cannot be removed by excretory organs. It gradually collects in the bodies of people living in areas where there is dense traffic, and may eventually begin to cause damage, especially to the brain.

Normally, pollutants disperse upwards as quickly as they are formed (Fig. 88.1A). But during a **temperature inversion**, when an upper zone of warm air traps cooler air at ground level (Fig. 88.1B),

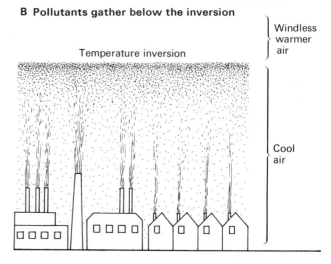

A Pollutants disperse upwards

Cooler air

Warm air

B Pollutants gather below the inversion

Windless warmer air

Temperature inversion

Cool air

Fig. 88.1 Comparison between normal conditions and those that lead to smog formation

fog and smoke gather, making thick yellow 'smog'. This happened in London in December 1952, and the death rate from bronchitis and similar diseases increased greatly.

Above : exhaust fume pollution

Right : exhaust fumes emitted by Concorde on take-off

Below : factory chimneys in Düsseldorf

Below right : decomposed stonework on Canterbury Cathedral

89

Water pollution

The main sources of water pollution are waste from houses, industry, agriculture, and spilled oil.

Sewage from houses can be made harmless by special treatment. In many countries, however, population growth has temporarily overloaded sewage works and so large amounts of untreated sewage are discharged into rivers and the sea. Sewage is eventually decomposed by bacteria, but in slow-moving rivers the bacteria quickly use up all the oxygen in the water. This kills fish, insects, and other aquatic life.

One of the main problems with industrial waste is that it often contains very poisonous, long-lasting pollutants such as compounds of cyanide, lead, mercury, and copper. These chemicals are dangerous even in small quantities because, when they are released into streams and rivers, they accumulate in fish and other aquatic creatures. They then spread throughout food webs to water birds, otters, and eventually to man.

The traditional agricultural method of spreading animal manure on land used to grow crops causes little or no pollution. In many modern farms, however, poultry, cattle, and pigs are kept in buildings and there is no other land on which to use the manure. Often this manure is discharged into local streams and rivers, where it decomposes and reduces oxygen levels in the same way as domestic sewage.

Other pollutants resulting from modern farming methods include chemical sprays which kill insects and fungi that attack crop plants. If these chemicals are washed into rivers and ponds they can spread through food webs in the same way as industrial waste.

Until oil supplies run out or man discovers an alternative and cleaner source of energy, there will be continual risk of oil spilling into the sea from tankers and offshore oil rigs. Spilled oil kills sea birds by poisoning them, and by sticking their feathers together so that they cannot fly. When it is washed ashore, oil kills all forms of life on rocks or in mud and sand. Fortunately, the sea contains bacteria which can decompose oil. After a period of time, bird colonies and shore life slowly return to normal.

Detergent polluting a river

A guillemot covered in oil

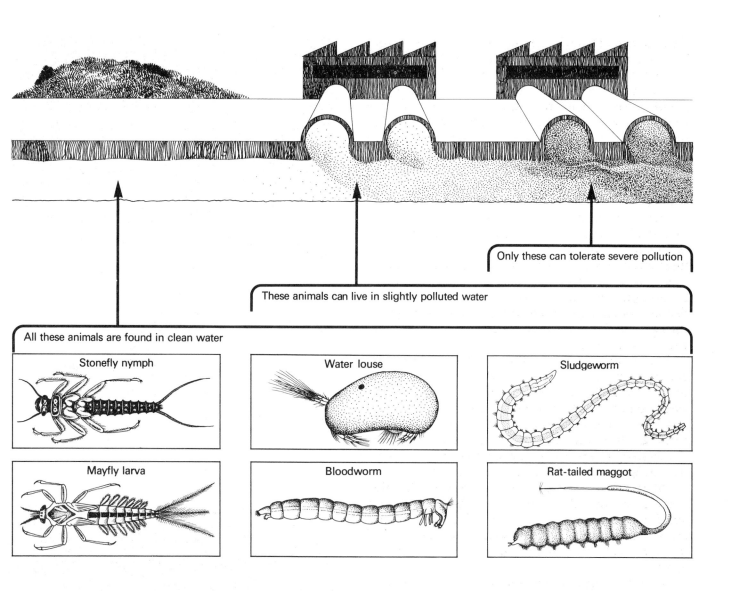

Only these can tolerate severe pollution

These animals can live in slightly polluted water

All these animals are found in clean water

Stonefly nymph	Water louse	Sludgeworm
Mayfly larva	Bloodworm	Rat-tailed maggot

Fig. 89.1 Diagram showing which animals can survive in polluted water

90

Conservation

Conservation means taking care of the world so that it continues to be a fit place for living things. It also means taking care of the living things themselves so that they do not become extinct.

Every time houses, shops, factories, roads, etc. are built, or a valley is flooded for a reservoir, or previously uncultivated land is ploughed for crops, the world becomes slightly less fit for living things – except humans. This is because all these activities rob wild animals and plants of places in which to live. Man poisons the world with his waste products, and destroys living things by eating them, and by hunting them, either for fur, skins, ivory, etc., or for sport.

These activities cannot be stopped completely, but they can be controlled in various ways. Areas of natural beauty and varied wild life can be made into **nature reserves**. Hunting and fishing are restricted inside reserves, and building is prohibited. With careful planning roads can be built to avoid areas rich in wild life, and cities can include areas of park land which, although they are not like wild country-side, do provide space where certain wild animals and plants can live.

Laws can be passed to protect living things. It can be made illegal to pick rare plants, hunt rare animals, or take eggs of rare birds. Hunting and fishing can be limited to periods of the year outside the main breeding season.

Conservation is not just the job of governments and local authorities. Everybody can help. People can dispose of their rubbish so that it does not litter the countryside. They can refuse to buy goods such as fur coats and crocodile skin hand-bags, which are unnecessary, and can only be produced by killing animals, some of which are becoming rare. If enough people complain about pollution, blood sports, or the waste of raw materials on unnecessary packaging and non-returnable containers, these things may eventually be stopped. But perhaps most important of all, people can teach their children that unless conservation is taken very seriously in the future, the world could become uninhabitable.

Fig. 90.1 What connection is there between each of these items and the problem of conservation?

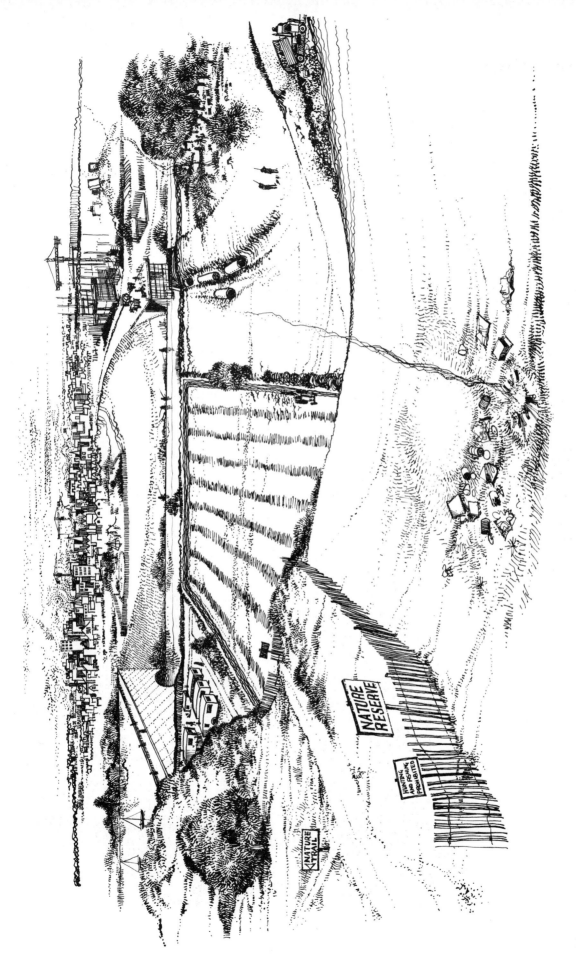

Fig. 90.2 Describe the problems of conservation shown in this drawing

Revision tests

Notes on the tests

Read these instructions carefully before attempting the tests.

1. Answers to most questions can be found by studying the text and illustrations in the units listed at the beginning of each test.

2. Some questions must be answered by drawing conclusions from information given in the text and drawings.

3. *Vocabulary tests*   These consist of a list of scientific words and a series of numbered phrases which describe the words in the list. The best way of doing these tests is to write out the numbered phrases and then, opposite each phrase, write the word which it describes. (*Note:* Many of the words are described more than once.)

4. *Comprehension tests*   These must be answered as fully as possible. Most of them, however, require only a brief paragraph or two of writing.

Test 1   Cells   (Units 10, 11, and 12)

Vocabulary test

nucleus, multicellular, organ, cell membrane, tissue, chromosomes, cytoplasm, vacuole, mitosis, cells, cellulose, chlorophyll, division of labour, zygote, chloroplasts.

1. Plant cells are enclosed in a layer of this substance.
2. These tiny objects contain chlorophyll and are found in plant cells.
3. They only become visible when a cell divides.
4. A semi-permeable membrane.
5. A word which means 'made of many cells'.
6. A group of specialized cells.
7. The part of a cell which contains chromosomes.
8. A large permanent space in the cytoplasm of a plant cell.
9. A type of cell division.
10. The 'building-blocks' of life.
11. Parts of a nucleus which contain 'instructions' for building an organism.
12. All the living material in a cell except the nucleus.
13. A space in cytoplasm filled with cell sap.
14. The green substance found in plants.
15. The extremely thin skin around a cell.
16. The advantage of this in multicellular organisms is that it leads to greater efficiency.
17. A group of specialized tissues.
18. The scientific name for a fertilized egg cell.
19. The heart and brain are two examples of this.

Comprehension test

1. Why are cells sometimes described as the 'building-blocks' of life?
2. Why is a cell membrane described as semi-permeable?
3. Describe the main differences between plant and animal cells.
4. Describe two animal tissues and one plant tissue.
5. Name as many organs in the human body as you can.
6. What is an organ system? Name three examples in the human body.
7. Why are the cells of multicellular organisms sometimes described as showing division of labour, and what are the advantages of division of labour?
8. Why can a human zygote develop only into a human and never into a dog or a horse?

Test 2   Skeletons, muscles, and movement
                                (Units 13, 14, and 15)
Vocabulary test

synovial fluid, exoskeleton, ligaments, insertion, endoskeleton, extensor muscle, chitin, origin, cartilage, antagonistic system, vertebrae, flexor muscle, intervertebral discs, tendons.

1. A muscle which straightens a limb.
2. Another name for gristle.
3. Arthropods have this type of skeleton.
4. Lubricating liquid of a joint.
5. The fixed anchorage point of a muscle.
6. A muscle which bends a limb.
7. Fibres which attach a muscle to the skeleton.
8. Pads of cartilage between the vertebrae.
9. The end of a muscle closest to the joint.
10. A muscle which straightens a limb.
11. The cuticle of an arthropod contains this substance.
12. Vertebrates have this type of skeleton.
13. A set of muscles which work in opposite directions.
14. The end of a muscle which moves during muscular contractions.
15. Bones which make up the backbone.
16. Tough fibres which hold bones together at the joints.
17. The extensor and flexor muscles of a joint are an example of this.

Comprehension test

1. What is the main difference between an exoskeleton and an endoskeleton?
2. Describe the main functions of the arthropod exoskeleton.
3. Describe the main functions of an endoskeleton.
4. What does the cuticle consist of in a crustacean and an insect?
5. What is the difference between the cuticle at the joints of an arthropod and elsewhere on the body?
6. Describe a typical slightly moveable joint.
7. Describe two features of a synovial joint which reduce friction at the point where the bones rub together.
8. Describe the main features of the following joints and name an example of each in the human skeleton: hinge joint, ball-and-socket joint, gliding joint.
9. Name the flexor and extensor muscles which move the arm at the elbow, and name the bones that each muscle is attached to. Which of these muscles provides the effort which bends the arm, which straightens it again, and where is the fulcrum about which the movement takes place?

Test 3   Movement in birds and fish
                                    (Units 16 and 17)
Vocabulary test

aerofoil, secondary feathers, upstroke, yawing, swim bladder, downstroke, rolling, primary feathers, pitching, lift.

1. The force produced by an aerofoil as it moves through the air.
2. The front edge of the wing is higher than its trailing edge during this stroke.
3. The feathers at the tip of a bird's wing.
4. During this stroke the feathers separate like the slats of a venetian blind.
5. The feathers attached to the part of a wing which has an aerofoil shape.

6. The front edge of the wing is lower than its trailing edge during this stroke.
7. The feathers which produce both lift and forward thrust when a bird flaps its wings.
8. The curved shape of a wing.
9. This stroke drives the bird forwards through the air.
10. 'Corkscrewing' swimming movements.
11. This organ enables a fish to float weightless in the water.
12. Sideways movements of a fish.
13. Pelvic and pectoral fins help prevent this movement.
14. Sharks do not have one.
15. Dorsal and ventral fins help prevent these two movements.
16. See-sawing movements.

Comprehension test
1. Name the two main features of a bird which enable it to overcome the force of gravity?
2. Describe how lift is produced by an aerofoil as it moves through the air.
3. Describe, in your own words, all that happens during the upstroke and downstroke of a bird's wing.
4. Describe the features of a bird's skeleton and general body shape which increase the efficiency of its flight.
5. (a) Which fin do most fish use to push themselves forwards through the water?
   (b) Where are the muscles which move this fin?
   (c) Describe how these muscles produce swimming movements.
6. What are the functions of the dorsal, ventral, pelvic, and pectoral fins?
7. How does a fish become weightless in water, and what is the advantage of this ability?

Test 4 Photosynthesis and plant mineral requirements   (Units 18 and 19)

Vocabulary test
chlorophyll, palisade layer, stomata, spongy layer, trace elements, glucose, phloem, chloroplasts, major elements, xylem vessels, mid-rib, mineral deficiency symptoms, veins, mesophyll.
1. The sugar produced by photosynthesis.
2. Microscopic objects containing chlorophyll.
3. Tissue through which sugar passes out of a leaf.
4. The layer of cells in a leaf where photosynthesis takes place.
5. Pores in leaves.
6. Nitrogen and phosphorus are examples.
7. The layer of cells between the upper and lower epidermis of a leaf.
8. Runs down the centre of most leaves.
9. The substance in leaves which absorbs light energy.
10. The uppermost region of the mesophyll.
11. Manganese and iron are examples.
12. These signs in plants tell farmers which minerals to add to their soil.

13. Found mostly on the under-surfaces of leaves.
14. Narrow tubes in which water passes through a plant.
15. Made of long cylindrical cells full of chloroplasts.
16. A branching network of xylem and phloem tubes in a leaf.
17. A layer of cells with large air spaces between them.

Comprehension test
1. What is the major difference in the way animals and plants obtain food?
2. What three things does a plant require to carry out photosynthesis?
3. Where and how do plants obtain these requirements?
4. What are xylem and phloem, and what are their functions?
5. Why does a plant die if it is left in the dark?
6. From what two substances does a plant produce glucose?
7. What is the waste product of photosynthesis?
8. What use do animals (and plants) make of this waste product?
9. What substances does a plant require to make proteins?
10. Where do plants obtain these substances?
11. What are 'mineral deficiency symptoms'? Describe some examples, and explain why farmers should study them carefully.

Test 5 Types of food   (Units 20 and 21)

Make a copy of Chart 1 (see opposite page). Put a tick under one or more of the headings at the right-hand side of the chart to show what the foods in the left-hand column are mainly made up of.

Test 6 Vitamins   (Unit 21)

Make a copy of Chart 2 (see opposite page). Put a tick under one or more of the headings at the right-hand side of the chart to show which vitamins are described in the left-hand column.

Test 7 Teeth   (Unit 23)

Vocabulary test
pulp cavity, carnivore, crown, incisors, cement, herbivore, canines, enamel, premolars, dentine, carnassials, root, omnivore, molars.
1. An animal that only eats plant food.
2. The upper set of these teeth is missing in sheep.
3. Long pointed teeth in carnivores.
4. The part of a tooth embedded in bone.
5. They have four cusps in humans.
6. Forms the hard surface of the crown.
7. An animal that eats plants and animals.
8. Soft bone-like substance in a tooth.

184

Chart 1

		Carbohydrates	Fats and oils	Proteins	Vitamins and minerals
1.	Baked potato				
2.	Fried potato				
3.	White bread and butter				
4.	Brown bread and butter				
5.	Steamed haddock				
6.	Fried cod				
7.	Orange juice				
8.	Raw egg				
9.	Ice cream				
10.	Egg, lettuce, and tomato salad				
11.	Fried pork chop with mashed potato				
12.	Cod-liver oil				
13.	Cheese and tomato sandwich				
14.	Fried bacon and egg				
15.	Toffee				
16.	Suet pudding				
17.	Strawberries and cream				
18.	Fried fish, chips, and peas				
19.	Milk chocolate				
21.	Milk				
22.	Fried rice				
23.	Steamed rice				
24.	Rice pudding				
25.	Cream cake				

Chart 2

	A	B₁	B₂	B₁₂	C	D	K
Found in liver							
Found in green vegetables							
Found in oranges and lemons							
Found in yeast							
Found in fish-liver oils							
Lack of it causes rickets							
Found in wholemeal bread							
Destroyed by cooking							
Found in fish							
Lack of it causes beri-beri							
Helps the body obtain energy from food							
Lack of it causes pernicious anaemia							
Needed to heal wounds							
Lack of it causes poor night vision							
Lack of it causes scurvy							
Lack of it may cause bones to bend							
Needed for good vision in dim light							
Made by bacteria in the digestive system							
Helps the body to form protein and fat							
Found in nuts, peas, and beans							
Lack of it causes paralysis of the limbs							
Needed for healthy bones and teeth							
Destroyed by mincing or grating food							

9. Chisel-shaped in humans.
10. Helps hold a tooth in place.
11. Flesh eater.
12. In between the molars and canines in humans.
13. Used to crack bones and cut flesh.
14. The space in the middle of a tooth.
15. Dagger-like teeth found in carnivores.
16. Immediately beneath the enamel.
17. The part of a tooth above the gum.

Comprehension test
1. (a) In what ways are the teeth of humans and dogs similar? In what ways do they differ?
   (b) What are the functions of each type of tooth in a dog's jaw?
2. (a) How does the shape of human incisors make them suitable for the job of biting?
   (b) How does the shape of human molars and premolars make them suitable for the job of chewing food into small pieces?
3. Why is it important to chew food thoroughly?
4. How does a sheep use its incisors to cut grass?
5. (a) Describe two differences between sheep and human molars.
   (b) How does the shape of sheep premolars and molars make them suitable for the job of grinding grass into pulp?

Test 8   Digestion and the digestive system
(Units 22, 24, and 25)

Vocabulary test
oesophagus, bolus, amylases, duodenum, lipases, sphincter, stomach, proteases, gall bladder, absorption, peristalsis, assimilation, bile, bile duct, ileum, anus, faeces, villi, digestion, large intestine.
1. Muscular contractions which move food through the digestive system.
2. Enzymes which digest proteins.
3. A ring of muscle.
4. Two regions where villi are found.
5. The process which makes food soluble.
6. Tube which carries food to the stomach.
7. Movement of digested food through the intestine wall into the blood.
8. Indigestible material.
9. Made in the liver.
10. Where water and salts are removed from food.
11. Where food is mixed with pepsin and hydrochloric acid.
12. Finger-like projections from the small intestine wall.
13. Ball of food.
14. Enzymes which digest starchy food.
15. Enzymes which digest fats and oils.
16. The part of the small intestine beyond the duodenum.
17. Bile is stored here.
18. Where faeces are ejected from the body.

19. The processes by which cells make use of digested food.
20. A tube which carries bile to the duodenum.
21. The region between the stomach and the ileum.
22. Changes fat into an emulsion.

Comprehension test
1. What is a digestive enzyme, and how does it carry out its function?
2. List the following processes in the order in which they occur in the body: absorption, assimilation, digestion.
3. What are the functions of saliva?
4. Describe what happens to food inside the stomach.
5. (a) Where is bile produced and stored?
   (b) What are its main functions?
6. What must carbohydrates, fats, and proteins be broken down into before they can be absorbed?
7. Which types of food do not have to be digested, and why is this so?
8. Describe the structure and functions of villi.

Test 9   Saprophytes, parasites, symbiosis and commensalism
(Units 27, 28, and 29)

Vocabulary test
sporangium, symbiosis, conjugation, bladderworm, saprophytes, primary host, parasites, mycelium, absorptive hyphae, commensalism, spores, secondary host, zygospore, hyphae.
1. Swelling at the tip of an upright hypha.
2. Bag of liquid containing a tapeworm head.
3. They live by digesting dead organisms.
4. This can only occur between 'plus' and 'minus' strains of *Mucor* hyphae.
5. This is always a vertebrate animal.
6. The collective term for all the hyphae in a mould colony.
7. They obtain food from the living bodies of other organisms.
8. Hyphae which digest and absorb food.
9. Carries an adult tapeworm in its intestine.
10. A relationship between two different organisms in which both benefit.
11. Fine, hollow threads.
12. Tiny oval objects containing many nuclei.
13. It carries bladderworms in its tissues.
14. Organs which produce asexual spores.
15. A relationship between two different organisms in which only one benefits.
16. They form inside sporangia.
17. Hyphae which grow down into food.
18. A spore produced by sexual reproduction.

Comprehension test
1. How does *Mucor* produce asexual spores, and how are these spores spread from place to place?

186

2. What is conjugation and how does it occur?
3. How does *Mucor* obtain food, and what is the technical word which describes organisms that feed in this way?
4. Describe the features of a tapeworm which enable it to survive in the intestine of its host.
5. What happens to a tapeworm egg when it is eaten by a pig?
6. How does the production of large numbers of eggs help tapeworms to complete their life cycle?
7. How do people become infected with tapeworms, and how can infection be prevented?
8. What benefits do bees and flowering plants obtain from the relationship they have with each other?
9. How do the organisms which make up a lichen benefit from their relationship?
10. The tick-bird lives on ticks which it removes from the skin of a rhinoceros. Is the relationship between the tick-bird and the rhinoceros symbiosis or commensalism?

Test 10   Transport in plants

(Units 30 and 31)

Vocabulary test

xylem vessels, phloem sieve tubes, xylem fibres, companion cells, root hairs, vascular bundle, cambium, osmosis, semi-permeable membrane, root pressure, transpiration, stomata, transpiration stream.
1. Divided into sections by sieve plates.
2. A strand made of xylem, cambium, and phloem.
3. They transport water and dissolved minerals.
4. A membrane which allows water to pass through but not certain dissolved substances.
5. They transport liquid from the roots to the leaves.
6. Pores in a leaf.
7. Diffusion of water from a weak to a strong solution through a semi-permeable membrane.
8. They transport sugar from the leaves to growing and storage areas.
9. Thick pointed fibres alongside xylem vessels.
10. A force which pushes water from root xylem into stem xylem.
11. The name for the flow of water and dissolved minerals from the roots to the leaves.
12. Situated alongside sieve tubes.
13. Cells which produce new xylem and phloem.
14. Loss of water from a plant by evaporation.
15. Strengthened by coils and rings of lignin.
16. These grow out of cells near the tip of a root.

Comprehension test
1. What is wood made of?
2. Describe the ways in which xylem vessels and phloem sieve tubes differ in both structure and function.
3. Describe how water gets from the soil into the xylem vessels of the root.

4. What is root pressure, and what part does it play in the movement of liquid through a plant?
5. What is transpiration, and what part does it play in the movement of liquid through a plant?
6. What is the transpiration stream, and which two forces produce it?
7. What are vascular bundles made of, and where are they found?
8. (a) What are stomata?
   (b) When, and how, do stomata open and close?

Test 11   Blood and its functions   (Unit 32)

Vocabulary test

red cells, platelets, phagocyte, haemoglobin, white cells, oxyhaemoglobin, blood clot, plasma.
1. The substance which gives red cells their colour.
2. Most of these have a large, irregularly shaped nucleus.
3. Tiny fragments of cells made in bone marrow.
4. Consists of more than 90% water, plus dissolved substances.
5. Slowly releases its oxygen to body cells.
6. The liquid part of blood.
7. Help blood to clot in wounds.
8. Disc-shaped cells concave on both sides.
9. Destroyed by the spleen.
10. Many move and change shape like *Amoeba*.
11. A substance formed as blood flows through vessels in the lungs.
12. A word which means 'cell-eater'.
13. Cells without a nucleus.
14. Blood which has been changed to jelly.
15. They live for about four months.
16. A chemical which enables red cells to transport oxygen.
17. This blocks wounds and prevents bleeding.

Comprehension test
1. (a) What is plasma?
   (b) List the substances transported around the body dissolved in plasma.
2. How do red and white cells differ in shape, appearance, size, and function?
3. (a) Where are red cells made?
   (b) Where are they destroyed?
4. What is oxyhaemoglobin, when is it formed, and what is its function?
5. What are phagocytic white cells, and how do they help fight infection?
6. (a) What are platelets?
   (b) Describe two ways in which the body would be in danger if there were no platelets in the blood.

Test 12   The heart, circulation, and tissue fluid   (Units 33, 34, and 35)

Vocabulary test

cardiac muscle, atria, auricles, ventricles, bicuspid,

tricuspid, semi-lunar, aorta, vena cava, pulmonary artery, pulmonary vein, arteries, capillaries, veins, tissue fluid, lymph gland.

1. The main artery.
2. Very narrow blood vessels.
3. The two lower chambers of the heart.
4. Valve between the right atrium and ventricle.
5. The main vein.
6. The heart is made of it.
7. Vessels which carry blood towards the heart.
8. The liquid which carries food and oxygen from the blood-stream to the cells.
9. Valve between the left atrium and ventricle.
10. The uppermost chambers of the heart (two possible answers).
11. Another name for 'atria'.
12. Valves situated at the points where blood flows out of the heart.
13. Gland with narrow channels which have white cells attached to the walls.
14. The thick-walled chambers of the heart.
15. Carries blood from the lungs to the heart.
16. These ensure that blood does not re-enter the ventricles after leaving the heart.
17. The thin-walled chambers of the heart (two possible answers).
18. Carries blood from the heart to the lungs.
19. Vessels which carry blood away from the heart.

Comprehension test
1. Describe, in your own words, the events which take place during one heart-beat.
2. What is the function of heart valves?
3. List the main differences between arteries, veins, and capillaries.
4. What is the pulse, and how is it formed?
5. How does the fact that many veins are situated inside large muscles help the blood to circulate?
6. What is the function of vein valves?
7. What is a capillary bed?
8. (a) How are tissue fluid and lymph formed?
   (b) What are their functions?

Test 13 Respiration and the human respiratory system

(Units 36, 37, 38, and 39)

Vocabulary test
respiration, internal respiration, external respiration, aerobic, anaerobic, fermentation, trachea, thoracic cavity, diaphragm, larynx, bronchi, bronchioles, alveoli, inspiration, expiration, oxygenated blood, deoxygenated blood, intercostal muscles.

1. The technical name for the windpipe.
2. Microscopic air-sacs.
3. The chemical breakdown of food to release energy for living things.
4. Respiration which does not use oxygen.

5. The scientific name for the voice box.
6. Breathing out.
7. Together these form the human respiratory surface.
8. Another name for cell respiration.
9. Blood which has absorbed oxygen.
10. Each is surrounded by a network of capillaries.
11. Very little energy is released by this type of respiration.
12. Dome-shaped sheet of muscle.
13. Muscles between the ribs.
14. Blood without oxygen.
15. The two branches of the trachea.
16. Anaerobic respiration in yeast.
17. Narrow branches of the bronchi.
18. Respiration which uses oxygen.
19. The scientific name for breathing.
20. This type of respiration produces alcohol.
21. The space in the chest which contains the lungs.
22. Breathing in.

Comprehension test
1. What is respiration, and what is the difference between respiration and breathing?
2. What chemicals are required for respiration to occur, and what waste products are released?
3. Describe the main differences between aerobic and anaerobic respiration.
4. Which microscopic organisms obtain energy by fermentation, and how does man make use of them?
5. When does anaerobic respiration occur (a) in plants, and (b) in the human body?
6. Explain, in your own words, what happens during inspiration and expiration.
7. What important things happen to air as it passes through the nasal passages?
8. What is gaseous exchange, and how does it occur in the human body?

Test 14 Respiratory organs in fish and insects

(Units 40 and 41)

Vocabulary test
operculum, gill bar, gill lamellae, tracheal system, tracheoles, spiracles.

1. There is usually a pair on each segment of an insect's body.
2. The region where gaseous exchange takes place in an insect's respiratory system.
3. Bone which supports a fish's gill.
4. Pores on the surface of an insect.
5. Flap of skin covering the gills in a fish.
6. Very narrow tubes deep in an insect's body.
7. Flaps of skin on a fish's gill covered with a network of capillaries.
8. The main part of an insect's respiratory system.

Comprehension test
1. Describe the structure of a fish gill.

2. Describe how a fish produces a continuous flow of water over its gills.
3. Why does a fish die if it is taken out of water?
4. What difference is there in the way oxygen and carbon dioxide are transported in insects and humans?
5. Describe the structure of an insect's tracheal system.
6. What is the function of the folds of cuticle around an insect's tracheal tubes?
7. Describe two ways in which air is drawn in and out of an insect's tracheal system.
8. What function do the air-sacs found in some insects have?

Test 15 Skin and temperature control
(Unit 42)

Vocabulary test
homoiothermic, poikilothermic, sweat, vasodilation, vasoconstriction, superficial capillaries, erector muscles, shivering, insulation.
1. These make hairs stand on end.
2. Tiny blood vessels just under the skin.
3. A word which describes animals whose temperature depends on their surroundings.
4. Enlargement of blood vessels.
5. A word which describes how fat helps reduce loss of heat from the body.
6. Contraction of blood vessels.
7. A word which describes animals which maintain a fairly constant body temperature.
8. Liquid which evaporates from the skin and cools the body.
9. It occurs in certain capillaries when the body is too hot.
10. They undergo vasodilation in hot weather.
11. A word which applies only to birds and mammals.
12. When this happens the skin loses a great deal of heat by radiation.
13. This occurs in certain capillaries when the body is too cold.
14. Jerky muscular movements which occur in cold weather.

Comprehension test
1. What is the difference between homoiothermic and poikilothermic animals?
2. Describe three ways in which the body can become overheated.
3. Why does vigorous exercise produce heat in the body?
4. How does the body prevent over-heating?
5. Water evaporates faster in hot, dry, windy weather than in hot, humid, windless weather. Why do we feel cooler on a hot, dry, windy day than on a hot, humid, windless one even if the temperature is the same on both days?
6. Describe three ways in which a constant body temperature is maintained when someone leaves a warm room and goes outside in cold weather.
7. Describe two ways in which hair reduces loss of heat from a mammal's body in cold weather.
8. Why do squirrels eat extra food in autumn?

Test 16 Excretion by the kidneys
(Unit 43)

Vocabulary test
urea, excretion, urine, urinary system, renal artery, renal vein, Bowman's capsule, glomerulus, glomerular filtrate, kidney tubules, reabsorption, urination, ureters, bladder.
1. Liquid which gathers in the bladder.
2. A vessel which carries blood away from a kidney.
3. A waste substance produced by the liver.
4. Produced when blood is filtered inside a Bowman's capsule.
5. A process which changes glomerular filtrate into urine.
6. Consists of the kidneys, ureters, and bladder.
7. Removal from the body of waste and unwanted substances.
8. Contains urea plus many useful substances.
9. Temporary store for urine.
10. A ball of inter-twined capillaries.
11. Removal of useful substances from glomerular filtrate.
12. Where reabsorption takes place.
13. A vessel which carries blood into a kidney.
14. Blood is filtered through its walls into a Bowman's capsule.
15. They carry urine from the kidneys to the bladder.
16. The liquid excreted by the kidneys.
17. Release of urine from the bladder.
18. Each contains a glomerulus.

Comprehension test
1. Name two substances which are excreted from the body, and the organs which excrete them.
2. List the parts of the urinary system.
3. Where, and how, is blood filtered in a kidney?
4. What is reabsorption and where does it take place?
5. What is the function of the bladder?

Test 17 Eyes and vision
(Unit 45)

Vocabulary test
sclerotic layer, cornea, conjunctiva, choroid, iris, pupil, lens, retina, suspensory ligaments, ciliary muscles, hypermetropia, aqueous and vitreous humours, optic nerve, fovea, blind spot, conjunctiva, myopia.
1. Made of nerve endings which are sensitive to light.
2. A layer of black substance inside the eyeball.
3. A disc of muscles with a hole in its centre.
4. These hold the lens in place.
5. Transparent substances which fill the eyeball.
6. Transparent skin which covers the front of the eyeball.

8. Conveys nerve impulses from the eye to the brain.
9. Small in bright light, and large in dim light.
10. A layer of tough white fibres in the eyeball.
11. Muscles make it fatter or thinner in shape as it focuses on near or distant objects.
12. Has the same position in the eye as a film has in a camera.
13. Prevents light from being reflected around inside the eyeball.
14. Muscles which control the shape of the lens.
15. Controls the amount of light entering the eye.
16. The part of the retina which is not sensitive to light.
17. The part of the retina responsible for colour vision.
18. The hole at the centre of the iris.
19. The part of the retina where blood vessels and nerves join the optic nerve.
20. The scientific name for long sight.
21. The scientific name for short sight.

Comprehension test
1. Describe the structure of the iris, its functions, and how it carries out these functions.
2. What shape is the lens when an eye is focused on a near object, and how is it made to form this shape?
3. What shape is the lens when an eye is focused on a distant object, and how is it made to form this shape?
4. What is the retina made of, and what is its function?
5. What is hypermetropia, what can cause it, and how can it be corrected?
6. What is myopia, what can cause it, and how can it be corrected?

Test 18   Ears, hearing, and balance
(Units 46 and 47)
Vocabulary test
pinna, ear drum, outer ear, middle ear, Eustachian tubes, ear ossicles, oval window, inner ear, semi-circular canals, cupula, ampulla, otoliths, cochlea.
1. Chain of three bones.
2. Two of them are upright and one is horizontal.
3. Made up of the pinna, ear canal, and ear drum.
4. Its funnel shape directs sounds towards the ear drum.
5. They lever the oval window in and out.
6. A series of tubular passages in the skull bones filled with perilymph.
7. An air-filled space behind the ear drum.
8. This collects sound waves travelling through the air.
9. Tiny pieces of chalk embedded in a jellyish substance.
10. A swelling at one end of a semi-circular canal.
11. These open during swallowing, letting air in or out of the middle ear.
12. Coiled like the shell of a snail.
13. A cone-shaped lump of jelly.
14. A sheet of skin and muscle which vibrates when sound waves reach it.
15. As vibrations move through it they stimulate sensory nerve endings which send impulses to the brain.
16. Contain sensory nerve endings embedded in a cupula.
17. Curved tubes concerned with the sense of balance.
18. These ensure air pressure is equal on both sides of the ear drum.
19. The ear ossicles lever it in and out, which causes vibrations to pass through the inner ear.
20. Tubes connecting the middle ear with the back of the mouth.

Comprehension test
1. Describe what happens to sound waves as they pass through the outer, middle, and inner ear.
2. (a) What are Eustachian tubes and what is their function?
   (b) How may hearing be affected if the Eustachian tubes become blocked?
3. (a) What is the function of the semi-circular canals?
   (b) What happens in the semi-circular canals when there is a sudden change in the direction of movement?
4. (a) What is the function of the utricles and saccules?
   (b) What happens in the utricles and saccules if the speed at which the body is moving changes?

Test 19   Tropisms
(Units 48 and 49)

Comprehension test
1. What are tropisms?
2. What is the difference between a tropic movement in a plant and a movement in an animal?
3. Name the tropic responses to light, gravity, and water.
4. (a) Describe the differences between a plant grown in the dark, in light from one side, and in light from above.
   (b) Which of these plants shows a phototropic response most clearly?
   (c) Study a plant grown in the dark (Fig. 48.1c), then describe other ways in which light affects plants.
5. What are the advantages to a plant of phototropic responses?
6. (a) What is a clinostat?
   (b) Why is it necessary to carry out Part B of the experiment illustrated in Fig. 48.2?
7. Describe how it is possible to conclude from the experiment illustrated in Figure 48.3 that only the tip of a plant responds to light? (Clue: pay special attention to Part A of the experiment.)
8. (a) What is the difference between positive and negative geotropism?
   (b) Which parts of the plants in Figures 49.1 and 49.3 show negative geotropism?
9. What does the experiment illustrated in Figure 49.1 show about the parts of a plant which bend during tropic responses?
10. Why must the experiment illustrated in Figure 49.2 take place in the dark?

Test 20   The nervous system and its functions   (Units 50 and 51)

Vocabulary test

co-ordination, central nervous system, spinal cord, nerve fibres, nerve impulses, cerebrum, cerebral hemispheres, cerebellum, medulla oblongata, sensory nerve cells, motor nerve cells, reflex action, synapse.

1. Thread-like extensions of a nerve cell.
2. The part of the brain which controls conscious behaviour.
3. 'Messages' which travel along nerves.
4. Cells which carry impulses out of the central nervous system.
5. Situated in a channel inside the backbone.
6. Receives impulses from the organs of balance.
7. Situated between the cerebrum and the medulla oblongata.
8. A tiny gap between nerve cells.
9. A word which describes how parts of the body work together in an orderly way.
10. The largest part of the human brain.
11. Co-ordinates muscles during walking.
12. Cells which carry nerve impulses from sense organs to the central nervous system.
13. An automatic response which requires no thought.
14. The two halves of the cerebrum.
15. Made up of the brain and spinal cord.
16. Controls the rates of breathing and heart-beat.
17. The response to a painful stimulus is an example.
18. Nerves are made up of thousands of these.
19. Made up of two cerebral hemispheres.

Comprehension test

1. (a) What does co-ordination mean?
   (b) Name some examples of co-ordination in the human body.
2. What is the cerebrum and what are its functions?
3. Why might someone with a damaged cerebellum be unable to ride a bicycle?
4. Which organs work faster when someone begins to run, and which part of the brain controls the rates at which these organs work?
5. What is a reflex action?
6. Describe the reflex actions which occur:
   (a) when dust blows into the eyes;
   (b) when bright light suddenly shines into the eyes;
   (c) when a person runs fast in hot weather;
   (d) when food accidentally enters the windpipe.

Test 21   The endocrine system   (Unit 52)

Vocabulary test

pituitary, thyroid, pancreas, testes, adrenal, hormones, insulin, oestrogen, thyroxine, diabetes, testosterone, adrenalin, ovaries.

1. Glands attached to the kidneys.
2. Part of the male reproductive system which produces hormones.
3. Produces digestive enzymes and hormones.
4. Gland at the base of the brain.
5. Gland which produces thyroxine.
6. Part of the female reproductive system which produces hormones.
7. Gland which produces adrenalin.
8. Gland attached to the windpipe.
9. These produce testosterone.
10. Organ situated just below the stomach.
11. Gland which produces hormones that control other endocrine glands.
12. Organ which produces insulin.
13. These produce oestrogen.
14. The chemicals produced by endocrine glands.
15. Disease caused by lack of insulin.
16. A female sex hormone.
17. A hormone which prepares the body for action.
18. A hormone which controls the rate of chemical reactions in the body.
19. A hormone which decreases the rate at which the liver releases glucose into the blood.
20. A male sex hormone.

Comprehension test

1. What are the main differences between the ways in which hormones and nerve impulses co-ordinate the body?
2. What are giantism and dwarfism and what causes them?
3. What is diabetes and what causes it?
4. What are the secondary sexual characteristics of human males and females, and which hormones control them?
5. In what ways does the pituitary gland affect the male and female reproductive systems?
6. Describe the effects of oestrogen on the female reproductive system.
7. Which hormone is produced during sudden emotional stress and what effects does it have on the body?

Test 22   Male and female reproductive systems   (Units 54, 55, and 56)

Vocabulary test

ova, sperms, ovary, testis, puberty, ovulation, scrotum, fallopian tube, uterus, vas deferens, cervix, vagina, vulva, menstrual cycle, semen, menstruation, epididymis.

1. This produces sperms.
2. The age at which sexual maturity is reached.
3. A sequence of events in females which lasts about 28 days.
4. Tube into which ova are released.
5. Scientific name for the womb.
6. Scientific name for the sperm duct.
7. Organ in which a baby develops.
8. This produces ova.
9. Storage area for sperms.

10. The scientific name for the release of an ovum from an ovary.
11. A tube between the uterus and vulva.
12. Bag in which the testes are situated.
13. The liquid which contains sperms.
14. External opening of the female reproductive system.
15. Age at which the production of sex cells begins.
16. Scientific name for eggs.
17. Ring of muscle at the base of the uterus.
18. Begins about 14 days after ovulation.
19. Male sex cells.
20. Female sex cells.

Comprehension test
1. What is an ovum, where do ova develop, and what happens to them immediately after they are released?
2. Describe the events which take place in the female reproductive system during one menstrual cycle.
4. Why is it almost impossible to predict a woman's fertile period?
5. What are the main differences between a sperm and an ovum?

Test 23　From ovulation to birth
(Units 57, 58, and 59)

Vocabulary test
follicle, fertilization, fertilization membrane, embryo, implantation, pregnancy, placenta, umbilical cord, amnion, foetus, afterbirth.
1. Fusion of a sperm with an ovum.
2. The period during which a female carries a developing baby.
3. The name for the ball of cells into which a fertilized ovum develops.
4. The bag of liquid which protects a developing baby.
5. The name for an embryo after eight weeks of development.
6. In human females it lasts nine months.
7. Thick skin around a fertilized ovum.
8. Bubble of liquid in which an ovum develops.
9. Disc-shaped organ through which a baby obtains nourishment and oxygen.
10. This develops immediately after the ovum is fertilized, preventing entry of more sperms.
11. It bursts at the moment of ovulation.
12. This contains the artery and vein which connect a baby with its placenta.
13. It is expelled from the womb after a baby has been born.
14. The process by which an embryo becomes embedded in the uterus wall.

Comprehension test
1. Describe the development of an ovum and its release from the ovary.

2. What is a follicle and what is its function after fertilization?
3. What changes occur in the uterus wall which prepare it to receive a fertilized ovum.
4. Describe fertilization, and the functions of the fertilization membrane.
5. What is implantation?
6. Describe the structure and functions of a placenta.
7. What is an amnion and what are its functions?
8. What causes labour pains?
9. Why is birth a dangerous moment for a baby?

Test 24　Reproduction in birds, frogs, fish, and insects
(Units 60, 61, 62, and 63)

Comprehension test
Birds
1. Describe some examples of courtship behaviour, and explain why it is necessary.
2. Describe the structure of a bird's egg, and the functions of its parts.
3. Describe how parent birds look after their young from the moment an egg is laid to the moment the young become independent.

Frogs
1. What are the main differences between a frog's egg and a bird's egg?
2. Describe how frogs' eggs are fertilized.
3. What is the main difference between fertilization in frogs and birds?
4. What difference is there between the feeding habits of a tadpole two weeks old, and a tadpole two months old?
5. When and how do a tadpole's internal gills develop?
6. (a) What is metamorphosis, and when does it occur in frogs?
(b) Describe metamorphosis in frogs.

Fish (stickleback)
1. What changes occur in the appearance of a male three-spined stickleback in early spring?
2. Describe how a male stickleback builds his nest.
3. Describe courtship, egg laying, and fertilization in sticklebacks.
4. Which parent looks after the young, how is this done, and for how long?

Insects
1. What is the difference between a nymph and a larva, and name an example of each?
2. (a) What is the difference between complete and incomplete metamorphosis?
(b) Describe an example of each type of metamorphosis.
3. What are the main differences between a larva and a pupa?
4. When is the cuticle shed during an insect's life cycle?

Test 25　Parts of a flower　　(Unit 64)

Vocabulary test

carpels, style, stigma, ovary, ovules, embryo sac, egg nucleus, stamens, filament, pollen sacs, anther, petals, corolla, nectary, sepals, calyx.

1. The female sex cell.
2. During pollination pollen grains land on this.
3. Situated in a ring around the carpels.
4. Consists of four pollen sacs.
5. The part of a carpel which contains ovules.
6. These develop into seeds.
7. Male reproductive organs.
8. These are situated in a ring outside the petals.
9. The part of a carpel which is fertilized.
10. Female reproductive organs.
11. The stalk which carries an anther.
12. Scientific name for all the petals of a flower.
13. Spaces inside an anther in which pollen grains grow.
14. The hollow base of a carpel.
15. Produces a liquid which attracts insects.
16. The part of a carpel immediately below a stigma.
17. Scientific name for all the sepals.
18. These may be brightly coloured and scented.
19. They form the protective outer covering of a bud.

Test 26　Pollination, fertilization, and development of seeds and fruits
　　(Units 65 and 66)

Vocabulary test

self-pollination, cross-pollination, unisexual, bisexual, pollen tube, micropyle, male gamete, female gamete, true fruit, false fruit.

1. Hole in an ovule through which the pollen tube passes.
2. Peas and beans are examples of this type of fruit.
3. The transfer of pollen from one flower to another on the same plant.
4. Flowers with both stamens and carpels.
5. A scientific name for the egg nucleus.
6. Apples and strawberries are examples of this type of fruit.
7. Flowers with either stamens or carpels but not both.
8. The transfer of pollen from anthers to stigmas in the same flower.
9. A tube through which a male gamete travels to a female gamete.
10. The transfer of pollen from one plant to another of the same species.
11. Grows out of a pollen grain.
12. The scientific name for the nucleus inside a pollen grain which fuses with the egg nucleus.

Comprehension test

1. Describe two ways in which self-pollination is prevented in flowers.
2. (a) Describe those features of a plantain flower

(Fig. 65.1) which show that it is wind-pollinated.
(b) Describe those features of a white dead-nettle (Fig. 65.2) which show that it is insect-pollinated.
(c) Describe some features of wind- and insect-pollinated flowers which are not found in either plantains or white dead-nettles.
3. Describe how self-pollination takes place in dandelion florets if cross-pollination fails (Fig. 65.3).
4. Describe how a male gamete reaches a female gamete in flowering plants.
5. Which part of a fertilized flower becomes a seed, and which part becomes the fruit?

Test 27　Dispersal and germination of seeds and fruits　　(Units 67 and 68)

Vocabulary test

testa, radicle, plumule, cotyledons, endosperm, germination, epigeal germination, coleoptile, hypogeal germination.

1. The scientific name for the stored food in cereal grains.
2. Germination in which the seed leaves are pulled above the ground.
3. Scientific name for the seed-coat.
4. Scientific name for the seed leaves.
5. Scientific name for the process by which a seed begins to grow.
6. Germination in which the seed leaves remain below ground.
7. Protects the growing plumule of a maize grain.
8. The scientific name for the root of an embryo plant.
9. In peas they are swollen with stored food.
10. Scientific name for the shoot of an embryo plant.
11. They shrivel and drop off when their stored food is used up.

Comprehension test

1. Why is it important that seeds do not germinate too close to the parent plant?
2. Describe one example of each of the following: wind dispersal, animal dispersal, and self-dispersal.
3. What must seeds be supplied with before they can germinate?
4. What happens to the testa, radicle, plumule, and cotyledons of a bean seed during germination?
5. (a) What is the difference between epigeal and hypogeal germination?
(b) Describe an example of hypogeal germination.

Test 28　Variation and heredity
　　(Units 70, 71, 72, and 73)

Vocabulary test

hereditary characteristics, acquired characteristics, discontinuous variation, continuous variation, hybrid, monohybrid cross, gene, genetics, factors.

1. The scientific study of heredity.
2. Variation in which there are no, or very few, intermediate forms.
3. Colour of hair and eyes are examples of these characteristics.
4. Scientific word for young produced by parents which differ in one or more hereditary characteristics.
5. Mendel's name for the microscopic 'particles' which he thought controlled hereditary characteristics.
6. Variation in which there are many intermediate forms.
7. Characteristics which can be inherited from parents.
8. Characteristics such as scars and knowledge are examples of this.
9. The modern name for what Mendel called a factor.
10. The name for a cross between parents which differ in only one way.
11. These characteristics are not inherited, but appear gradually and often by chance during a person's lifetime.

Comprehension test
1. Describe some examples of variations which occur in animals and plants. (Think of some which are not mentioned in Unit 70.)
2. Describe some examples of hereditary and acquired characteristics, besides those mentioned in Unit 70.
3. What is the difference between continuous and discontinuous variation?
4. Why was Mendel's choice of garden peas for his experiments a wise one?
5. What are pure lines?
6. (a) What is a monohybrid cross?
   (b) How did Mendel perform such a cross on garden peas?
7. Why did Mendel call tallness in pea plants dominant, and dwarfness recessive?
8. (a) Describe what Mendel meant by factors.
   (b) Why did Mendel believe that factors work together in pairs?
   (c) What evidence is there that factors are part of chromosomes?

Test 29   Evolution
(Units 74, 75, 76, 77 and 78)

Comprehension test
1. (a) What are fossils and how are they formed?
   (b) Why are fossils sometimes called the 'record of the rocks'?
2. What evidence is there from the study of embryology that humans may have evolved from a fish-like ancestor?
3. What evidence is there that horses, birds, whales, and humans all evolved from the same ancestor?
4. What do the theory of evolution and the theory of natural selection try to explain?

5. (a) What is the 'struggle for survival' and what evidence is there that such a struggle exists in nature?
   (b) How does the struggle for survival lead to the 'survival of the fittest'?
   (c) What is the connection between the survival of the fittest and natural selection, and how do these lead to evolutionary change?
6. (a) What is artificial selection?
   (b) How do the results of artificial selection support Darwin's theory of natural selection?
7. (a) Explain where, and why, dark-winged peppered moths survive better than the pale variety.
   (b) Explain where, and why, the pale variety of peppered moth survives better than the dark-winged variety.
   (c) How do these facts illustrate the theory of natural selection?
8. The Galapagos Islands are populated with finch-like birds found nowhere else in the world. What is Darwin's explanation of where these birds came from?

Test 30   Disease and the body's defences against it
(Units 79, 80, 81, and 82)

Vocabulary test
germs, vaccines, opsonins, pathogenic, antibody, vector, toxins, lysins, agglutinins, pathology, antitoxins, contagious, antigens.
1. The common name for bacteria and viruses.
2. Scientific word describing parasites which harm their host.
3. The study of diseases.
4. Scientific name for poisons produced by germs.
5. Scientific name for infection spread by direct contact.
6. An animal which spreads infection.
7. A chemical produced by the body which kills germs and their toxins.
8. Substances which stimulate the body to produce antibodies.
9. They appear to make germs more 'appetising' to phagocytes.
10. Antibodies which dissolve germs.
11. Antibodies which stick germs together.
12. Antibodies which make toxins harmless.
13. They either contain dead germs, or harmless germs similar to types which cause serious diseases.

Comprehension test
1. Name some diseases caused by pathogenic bacteria.
2. (a) How is infection spread by coughing and sneezing, food and water, coins and doorknobs, and animals?
   (b) Name some diseases spread in each of these ways.
3. Describe features of the respiratory system, skin, eyes, and digestive system which act as defences against infection.

4. (a) What is a blood clot, and how does it help prevent infection?
(b) How do phagocytes help prevent infection?
5. (a) What are antibodies, and when are they produced?
(b) Describe some types of antibody.
6. How are antibodies made, and how are these man-made antibodies used to fight germs?
7. What is a vaccine and what effect does it have when injected into the body?
8. Write an essay called 'The dangers of smoking'.

Test 31   Soil (Unit 83)

Comprehension test
1. (a) Where does soil humus come from?
(b) How does humus lead to the formation of soil crumbs?
(c) How do soil crumbs and humus help make a soil fertile?
2. Describe the ways in which soil rock particles (sand, silt, and clay) are formed.
3. (a) What is loam?
(b) Why is loam described as the most fertile type of soil?
4. (a) What are the differences between sandy soil and loam?
(b) Why is sandy soil less fertile than loam?
(c) How can the fertility of sandy soil be improved?
5. (a) What are the differences between clay soil and loam?
(b) Why are clay soils less fertile than loam?
(c) How can the fertility of clay soil be improved?

Test 32   Carbon and nitrogen cycles
(Units 84 and 85)

Comprehension test
1. (a) Name the materials that are constantly used by living things?
(b) Why are these materials never completely used up?
(c) Why are these processes called the 'balance of nature'?

Carbon cycle
2. (a) Name the process which uses carbon dioxide?
(b) Why does this process occur only during the day-time?
(c) What process in living organisms produces carbon dioxide?
(d) How is carbon dioxide produced when dead organisms decay?

(e) Where does the carbon come from which is contained in the carbon dioxide released during combustion of wood, coal, and oil?

Nitrogen cycle
3. (a) Why is nitrogen essential for life?
(b) How do plants and animals obtain nitrogen?
4. (a) How does lightning lead to the formation of nitrates in soil?
(b) Why is soil improved by growing peas, beans, or clover?
(c) How are soil nitrates formed by the decomposition of dead organisms?
5. Why does soil lose nitrates during a flood?

Test 33   Food chains, food webs, and the ecosystem (Units 86 and 87)

Vocabulary test
food chain, producers, consumers, pyramid of numbers, food web, community, habitat, ecology, scavengers, decomposers.
1. Rock-pools, marshes, and forests are examples of this.
2. Food chains always begin with these.
3. The organisms in a pond are an example.
4. They eat dead organisms.
5. The number of organisms in this decreases from the base to the top.
6. Energy in the form of food moves along it from one organism to the next.
7. Herbivores and carnivores are examples.
8. The scientific study of ecosystems.
9. Saprophytic fungi and bacteria are examples.
10. They cause dead organisms to decay.
11. They are given this name because they either eat plants, or they eat animals which eat plants.
12. Crows and maggots are examples.

Comprehension test
1. Explain why all living things depend on sunlight.
2. Arrange the following organisms in the order in which they would occur in a food chain.
(a) tadpole, water flea, pond weed, pike.
(b) caterpillar, hawk, cabbage, starling.
(c) robin, rose bush, ladybird, greenfly, kestrel.
3. What is a food web?
4. (a) What are producers, and why do all food chains begin with producer organisms?
(b) Why are consumers given this name?
(c) Name some scavengers and decomposers, and describe the part they play in a food web.
5. (a) What is a community?
(b) What is a habitat?
(c) Name some organisms you would expect to find in the following habitats: a pond; a hedgerow.
6. What makes up an ecosystem?
7. What do ecologists do?

Test 34 Pollution and conservation
(Units 88, 89, and 90)

Comprehension test

Pollution

1. When can a substance be described as a pollutant?
2. List some of the pollutants which are produced by power stations, factories, houses, and motorcars.
3. Describe some harmful effects to plants and animals, including humans, of:
   (a) burning coal and oil in power stations;
   (b) fumes from cars and buses;
   (c) sewage;
   (d) crop sprays.
4. What is a temperature inversion and in what way does it increase the effects of pollution?
5. (a) What are the main sources of oil pollution?
   (b) What are the main harmful effects of oil pollution on living things?
   (c) How is oil pollution dealt with by man, and what natural process eventually removes it?

Conservation

6. What is conservation?
7. Describe some of the ways in which man makes the world less fit for wild animals and plants.
8. Describe some ways in which harm to wildlife can be reduced.
9. How can ordinary people help conserve wildlife?

Glossary

This glossary gives brief definitions of some of the most important scientific words used in the text. A word in *italics* within a definition can be found elsewhere in the glossary.

Absorption   The movement of digested (soluble) food through the walls of the intestine into the blood-stream.

Accommodation   Alterations to the focus of the eye. In man this is done by changing the shape of the lens.

Adrenalin   A *hormone* secreted by the adrenal glands. It prepares the body for instant action by increasing the rate of heart-beat, blood pressure, and blood sugar level.

Aerobic respiration   A type of *respiration* in which oxygen is consumed.

Aerofoil   The shape of a bird's wing as seen in cross-section. It produces lift as the bird moves through the air.

Agglutinin   A type of *antibody* which sticks bacteria together in clumps, thereby hindering their ability to reproduce.

Alimentary canal   The intestine. A tube running from mouth to anus inside which *digestion* and *absorption* take place.

Alluvial soil   Soil made up of particles which have been carried by rivers or glaciers to their present position.

Alveoli   Bubble-like air pockets at the ends of the air passages in the lungs. They are surrounded by blood vessels and are concerned with *gaseous exchange*.

Amino acids   The chemical units which make up *protein* molecules. These separate from one another when *protein* is digested.

Amnion   The liquid-filled sac that surrounds and protects the *embryos* of reptiles, birds, and mammals.

Ampulla   The swollen portion of a *semi-circular canal*. It contains sensory hairs and a *cupula*, and detects changes in the direction of movement.

Amylase   A type of *enzyme* which digests *carbohydrates*.

Anaerobic respiration   A type of *respiration* in which oxygen is not consumed.

Antagonistic muscle system   Two sets of muscles which oppose each other at each side of a joint. One set bends the joint (flexor muscles) and the other straightens it (extensor muscles).

Antennae   Long narrow sense organs on the heads of insects, containing nerve endings responsible for senses of touch, taste, smell, humidity, and temperature.

Anther   The part of a *stamen* which produces and later releases *pollen*.

Antibodies   Chemicals made by the body in response to *parasites* and harmful substances called *antigens*. They destroy or neutralize antigens.

Antigens   Bacteria, viruses, or harmful substances in the body which stimulate the production of *antibodies*.

Antitoxin   A type of *antibody* which neutralizes poisonous substances, particularly those produced by germs.

Aorta   The main *artery* of the body.

Aqueous humour   The liquid that fills the front chamber of the eye between the lens and the *cornea*.

Artery   Blood vessel that carries blood away from the heart.

Asexual reproduction   Reproduction involving one parent, without the fusion of *sex cells*.

Atrium (auricle)   One of the thin-walled upper chambers of the heart. It receives blood from the *veins*.

Bile   A greenish-yellow liquid made in the liver and passed into the *duodenum*, where its main function is to aid in the *digestion* of fats.

Binary fission   A type of *asexual reproduction* in which the organism divides into two parts. Typical of *Amoeba*.

Bladderworm   A stage in the life cycle of a tapeworm which lives in the muscles of the secondary *host*. It consists of a fluid-filled bladder containing a tapeworm head.

Blind spot   Point at which the optic nerve leaves the *retina* of the eye. It is not sensitive to light.

Bowman's capsule   A cup-shaped structure in a kidney 0.1 mm in diameter in man. It contains a *glomerulus* and leads to a *kidney tubule*.

Bronchial tree   The branching air passages of the lungs starting with the single *trachea* and ending in millions of fine bronchioles.

Budding   A type of *asexual reproduction* in which young develop as outgrowths from the parent's body.

Bulb   An organ of *vegetative reproduction* consisting of a very short stem surrounded by thick leaves swollen with water and stored food (e.g. daffodil bulb).

Caecum   A blind-ended branch of the intestine at the junction between the *ileum* and *colon*.

Calyx   All the *sepals* of a flower.

Canine teeth   Conical, dagger-like teeth at each side of the jaws of some mammals. These teeth are well developed in *carnivores*, which use them to kill their prey and tear meat from their bones.

Capillary Narrow, thin-walled blood vessel conveying blood from *arteries* to *veins*. The main exchanges of gases and dissolved substances between the blood and body cells take place through capillary walls.

Carbohydrates Sugary and starchy foods.

Carbon cycle The continuous circulation of carbon atoms between carbon dioxide in the air and the bodies of living organisms.

Carnivore Flesh-eating animal, e.g. lion, tiger.

Carpel Female sex organ of a flower.

Cell A unit of living matter, consisting of a *nucleus*, *cytoplasm*, and a *cell membrane*.

Cell membrane The *semi-permeable* membrane which forms the outer surface of all cells.

Cellulose A carbohydrate made up of long fibres. It forms the rigid cell wall which surrounds all plant cells.

Central nervous system Brain and spinal cord.

Cerebellum The part of the brain which controls balance and muscular co-ordination.

Cerebral cortex *Grey matter* which forms the outer layer of the *cerebral hemispheres*. It controls voluntary movements, and is concerned with memory, thinking, and learning.

Cerebral hemispheres Two swellings at the top of the brain. These form the largest region of the human brain and are concerned with conscious sensation, learning, and memory.

Chlorophyll Green substance in plants. It absorbs light energy for use in *photosynthesis*.

Chloroplasts Microscopic objects in plant *cells*, containing *chlorophyll*.

Chromosomes Rod-like structures visible in the *nucleus* of a *cell* during cell division. They contain the *hereditary information* of the *cell*.

Cilia Minute hair-like projections from the surface of certain *cells*, e.g. *Paramecium*. Cilia flick back and forth causing the surrounding fluids to move.

Ciliary muscles Muscles in the eye which change the shape of the lens during focusing.

Cochlea A spiral tube within the inner ear, containing sensory nerve endings which respond to sound vibrations.

Coleoptile Sheath-like protective covering over the first-formed leaves of grasses and other cereals.

Collecting tubule A tube within a kidney which collects *urine* from the *kidney tubules* and passes it into the *ureter*.

Colon Part of the large intestine. Its function is to absorb water and mineral salts from *faeces*.

Commensalism A close association between two different organisms in which one of them benefits.

Community A group of interdependent organisms that share a particular environment.

Companion cells Elongated cells situated alongside a sieve tube of *phloem*.

Conjugation A type of *sexual reproduction* in which cells join in pairs and exchange *hereditary information*.

Conjunctiva Transparent skin that covers and protects the front of the eye.

Conservation Taking care of the world so that it continues to be a fit place for living things.

Consumers Organisms in a *food-chain* which live by consuming (eating) other organisms.

Co-ordination Control of the body so that its tissues and organs work together at the correct speed and in the correct sequence, thereby serving the body as a whole.

Corm An organ of *vegetative reproduction* consisting of a short stem swollen with stored food.

Cornea Transparent circular window at the front of the eye.

Corolla All the petals of a flower.

Cupula A cone of jelly attached to sensory hair cells in the *ampulla* of a *semi-circular canal*.

Cytoplasm All the contents of a *cell* except its *nucleus*.

Denitrifying bacteria *Anaerobic* bacteria in soil which break down *nitrates* into nitrogen and oxygen.

Dentine A substance similar to bone which forms the inner part of a tooth, beneath the *enamel*.

Dermis The layer of skin beneath the *epidermis*, consisting of connective tissue, blood vessels, nerves, hair roots, and cells filled with fat.

Diaphragm Dome-shaped sheet of muscle at the base of the *thorax*. It is part of the mechanism which causes breathing in and out.

Diffusion Movement of molecules of liquids and gases from regions where they are highly concentrated to regions where they are less concentrated until they are equally distributed.

Digestion The process by which food is made soluble by the action of digestive juices (*enzymes*).

Dominant characteristic A characteristic of parents which appears in their young when crossed with a *recessive characteristic*.

Duodenum The part of the digestive system between the *stomach* and the *ileum*.

Ear drum A membrane of skin and muscle situated at the bottom of the ear canal of the outer ear. It is vibrated by sound waves in the air and transmits these vibrations to the ear *ossicles*.

Egg nucleus A *nucleus* inside the *ovule* of a plant which fuses with a male nucleus from a *pollen* grain.

Embryo The stage of development between the fertilized egg (*zygote*) and the newly formed organism.

Embryo sac Part of the *ovule* of a plant which contains the *egg nucleus*.

Enamel The extremely hard, white substance which forms the outer surface of a tooth.

Encystment A process in *Amoeba* and other protozoa in which the organism becomes enclosed in a hard outer covering. In this state it can survive conditions which would kill it in the free-living stage.

Endocrine system A system of organs which produce hormones.

Endoskeleton A skeleton which forms inside the body of an organism.

Enzymes Protein substances which control the rate of chemical reactions in *cells*, generally speeding them up.

Epidermis The outer layer of *cells* in an animal or plant.

Eustachian tube An air-filled tube running from the back of the mouth to the middle ear. It permits air pressure to be equalized on either side of the *ear drum*.

Evolution The sequence of gradual changes over millions of years in which new *species* may be produced.

Excretion Removal from the body of waste substances and substances which are in excess of the body's requirements.

Exoskeleton A skeleton which forms on the outside of an organism.

F₁ generation The first filial generation. Organisms produced by crossing animals or plants which form the starting point of a genetic experiment.

F₂ generation Organisms produced by crossing or selfing members of an F₁ generation.

Faeces The indigestible material which remains in the *colon* after *digestion* has taken place.

Fermentation The breakdown of sugar by organisms such as yeast and bacteria which takes place under *anaerobic* conditions.

Fertilization The fusion of male and female *sex cells* during *sexual reproduction*. It results in the formation of a *zygote*, which becomes a new organism.

Foetus The *embryo* of a mammal at the stage of development in which the main features are visible.

Food chain A number of organisms which feed on each other. The chain always begins with *producers* (green plants) and the 'links' in the chain are *consumers* (mainly animals).

Food web A number of interconnected *food chains*.

Gall bladder A small bladder inside the liver in which *bile* is stored.

Gaseous exchange The process by which an organism absorbs oxygen from the air in exchange for carbon dioxide, which is released into the air. This takes place in respiratory organs such as lungs.

Gene Part of a *chromosome* which controls the appearance of a set of *hereditary characteristics*.

Genetics The scientific study of *genes*.

Geotropism Growth movement of a plant in response to gravity.

Gland A collection of cells which manufactures and releases into the body useful substances such as *enzymes* or *hormones*.

Glomerular filtrate Fluid which results from the filtration of blood in *Bowman's capsules*. It consists of *urine* and many useful substances such as glucose.

Glomerulus A group of *capillaries* inside a *Bowman's capsule* in a kidney. Blood is filtered as it passes through the glomerulus and Bowman's capsule walls into the *kidney tubule*.

Grey matter Nervous tissue in the brain and spinal cord.

Guard cells Crescent-shaped cells in the *epidermis* of plants which surround and control the diameter of stomatal pores (*stomata*).

Habitat A region of an environment containing its own particular *community* of organisms, e.g. marsh, sand dune, pond.

Haemoglobin Red substance in *red blood cells*. It combines with oxygen and transports it to the tissues.

Hepatic portal vein Vessel in which blood containing absorbed food is carried from the intestine to the liver.

Herbivore An animal which eats only plants, e.g. horse, sheep.

Hereditary characteristics Those characteristics which an organism inherits from its parents (e.g. eye colour).

Hermaphrodite An organism which possesses both male and female reproductive organs.

Homoiothermic (organisms) Those which are capable of maintaining a constant body temperature.

Hormone A chemical produced in small amounts which helps to co-ordinate processes such as growth and reproduction.

Host An organism in or on which a *parasite* lives.

Humus Part of the soil which consists of the decayed remains of plants and animals.

Hydrotropism A growth movement in plants in response to water.

Hyphae Fine hollow threads which make up the body of many fungi.

Ileum The region of the digestive system between the *duodenum* and *colon*, where *digestion* is completed and *absorption* takes place.

Imago Fully developed adult insect.

Immunity The ability of an organism to resist infection from *parasites*.

Incisors Chisel-shaped teeth at the front of the jaws.

Insulin A *hormone* produced in the *pancreas*. It helps to control the amount of sugar in the blood.

Intercostal muscles Muscles between the ribs that raise the rib cage during inspiration (breathing in).

Iris The coloured part of the eye, consisting of muscles which alter the size of the *pupil* and control the amount of light entering the eye.

Larva An early stage in the life cycle of certain organisms which bears little or no resemblance to the adult (e.g. caterpillar).

Ligament Band of fibres around a joint in the skeleton which holds the bones in place and helps prevent dislocation.

Lignin A hard rigid substance which forms in the walls of *cells* which make up *xylem tissue*.

Lipase An *enzyme* which digests fats and oils.

Loam A soil containing sand, clay, and *humus* in ideal proportions for healthy plant growth.

Lymph A liquid derived from *tissue fluid* after it has passed between the cells of the body and drained into the *lymphatic system*.

Lymphatic system A system of tubes that transport *lymph* from the tissues to the circulatory system.

Lymph node (gland) Part of the *lymphatic system* containing *phagocytes*, which remove germs and dead cells from the *lymph*.

Malnutrition The harmful effects on the body of either too much, or too little food.

Menstruation Breakdown and removal from the body of the lining of the *uterus*. This occurs if an *ovum* has not been fertilized.

Metabolism All the chemical and physical processes necessary for life.

Metamorphosis The series of changes by which a *larva* becomes an adult organism.

Mid-rib The rigid rib in the centre of a leaf, containing *xylem* and *phloem*.

Mitosis A type of cell division resulting in cells with the same number of *chromosomes* as the parent cell.

Molars Large teeth with four cusps, situated at the back of the jaw. They are used to crush and grind food into small pieces.

Monohybrid cross Cross between organisms which show contrasting variations of only one characteristic.

Myopia Short-sightedness. Usually results from an abnormally elongated eyeball.

Natural selection A theory proposed by Charles Darwin, suggesting how evolutionary change could have occurred.

Nectary An organ in a flower that produces the sugary fluid nectar, which aids *pollination* by attracting insects.

Nitrates Substances made of nitrogen and oxygen. Vital for healthy growth of plants.

Nitrifying bacteria Bacteria in soil which convert the decaying remains of organisms into soil *nitrates*.

Nitrogen-fixing bacteria Bacteria in soil and *root nodules* which convert nitrogen in the air into soil *nitrates*.

Nucleus Part of a cell which contains *chromosomes*. It controls cell *metabolism* and division.

Nymph An early stage in the life cycle of certain insects that resembles the adult except that it is smaller, usually wingless, and sexually immature.

Oesophagus Tube through which food passes from the mouth to the *stomach*.

Oestrogen Female sex *hormone*, controlling conditions in the *uterus* before and during *pregnancy*.

Omnivore An animal that eats both animals and plants, (e.g. man).

Opsonins *Antibodies* which combine with chemicals on the surface of bacteria, making them more likely to be attacked by *phagocytes*.

Organ A structure consisting of several *tissues* which work together in performing a particular function, (e.g. the heart).

Origin (of a muscle) The anchorage point of a muscle, i.e. the end which does not move during contraction.

Osmosis *Diffusion* of water molecules through a semi-*permeable membrane* from a weak to a strong solution.

Ossicles (of the ear) Three tiny bones in the middle ear which transmit vibrations from the *ear drum* to the inner ear.

Otoliths Minute grains of chalk embedded in blobs of jelly attached to sensory hairs in the *utricles* and saccules of the inner ear. They are displaced by body movements, thereby causing the hair cells to send impulses to the brain.

Ova (singular **ovum**) Female *sex cells* of animals.

Ovary An organ which produces female *sex cells* (e.g. eggs).

Ovulation The release of an *ovum* from an *ovary*.

Ovule The part of a *carpel* which develops into a seed after *fertilization*.

Oxygen debt This occurs in muscle tissue during strenuous exercise, when oxygen is consumed faster than it can be supplied by the blood.

Oxyhaemoglobin *Haemoglobin* which has combined with oxygen in the *red blood cells*.

Palisade mesophyll A layer of cylindrical *cells* at right angles to the upper *epidermis* of leaves. They contain more *chlorophyll* than other plant cells, and are the main cells concerned with *photosynthesis* in plants.

Pancreas An organ situated between the *stomach* and the *duodenum*, producing digestive *enzymes*.

Parasite An organism that obtains food from the living body of another organism called the *host*.

Pathology The scientific study of the effects on the body of disease organisms.

Pepsin An *enzyme* produced by the *stomach* which begins the the *digestion* of *proteins*.

Peristalsis Wave-like contractions of tubular organs such as the digestive system, that move the contents of the tube in one direction.

Petiole Leaf stalk.

Phagocytes *White blood cells* that engulf and digest germs.

Phloem A plant tissue that transports the products of *photosynthesis* from the leaves to the growing points and food storage organs. It consists mainly of *sieve tubes* and *companion cells*.

Photosynthesis The process by which plants use light energy trapped by *chlorophyll* to form sugar out of carbon dioxide and water.

Placenta The organ through which the *foetus* of a mammal obtains food and oxygen from its mother's blood, and passes waste substances into the mother's blood.

Plasma The liquid part of blood.

Plasmolysis The shrinkage of *cell cytoplasm* owing to loss of water by *osmosis*.

Platelets Particles in the blood which are concerned with clot formation in wounds.

Pleural cavity The fluid-filled space between the outer surface of the lungs and the inner surface of the rib cage.

Poikilothermic (organisms) Those which cannot maintain a constant body temperature, but vary according to the temperature of their surroundings.

Pollen Male *sex cells* of flowering plants.

Pollination Transfer of *pollen* grains from *stamens* to *stigmas*.

Pollutant A substance which is harmful to living things.

Pregnancy The period during which a female mammal carries a developing *embryo* in her *uterus*.

Presbyopia Old sight. A condition resulting from old age in which the lens loses its ability to change shape during focusing.

Producers Organisms such as green plants that produce food. Producers form the starting points of *food chains*.

Protease An *enzyme* which *digests protein*.

Proteins The main body-building foods such as meat, eggs, and fish.

Pseudopodia Projections from the cytoplasm of certain *cells*, e.g. *Amoeba*. They are used in feeding and movement.

Puberty The stage of development at which men and women become sexually mature, i.e. able to reproduce.

Pupa A stage in the life cycle of an insect during which *metamorphosis* takes place (e.g. butterfly chrysalis).

Receptacle The surface at the top of a flower stalk to which all the parts of a flower are attached.

Recessive characteristic One that does not appear in the young, when crossed with a *dominant characteristic*.

Rectum The last part of the digestive system.

Red blood cells Disc-shaped cells containing *haemoglobin*, which transports oxygen from the lungs to the body tissues.

Reflex A response that does not have to be learned, and which occurs very quickly without conscious thought, e.g. pulling the hand away from a hot object.

Respiration A series of chemical reactions which release energy from food.

Response An activity in the body that results from a *stimulus*.

Retina A layer of light-sensitive cells at the back of the eye on which images are projected.

Rhizome A horizontal underground stem filled with stored food. It enables a plant to survive the winter and make new growth in the spring, and produces new plants from branches which grow from underground buds.

Root hairs Hair-like outgrowths from single *cells* in the *epidermis* of a root in a zone near the root apex.

Root nodules Swellings on the roots of certain leguminous plants (such as peas and clover), containing *nitrogen-fixing bacteria*.

Root pressure Pressure causing water to pass up the *xylem* from the living *cells* of the root.

Saliva Fluid produced and released into the mouth by three pairs of salivary glands, in response to food. It contains an *enzyme* which begins the *digestion* of cooked starch.

Saprophytes (saprotrophs) Organisms which feed on the dead remains of animals and plants, by releasing *enzymes* that *digest* the food externally, reducing it to a liquid which is absorbed into the saprophyte body, e.g. certain bacteria and fungi.

Sebaceous gland A *gland* in the hair follicles of the skin. It *secretes* an oily substance called sebum, which makes skin supple, waterproof, and mildly antiseptic.

Secretion The production by *glands* of substances such as *enzymes* which are useful to the body.

Sedentary soil Soil which overlies the rocks from which it was formed.

Semen Fluid produced by the *testes* of mammals. It consists of *sperms* and chemicals which nourish them and stimulate their swimming movements.

Semi-circular canals Three curved tubes in each inner ear. They contain fluid and sensory hair cells attached to the *cupula*. They respond to changes in the direction of movement.

Sepals Leaf-like structures at the outer region of a flower. They cover and protect the flower in bud.

Sex cells Cells which fuse together at *fertilization* (e.g. sperms, eggs, ovules, pollen grains). They form a *zygote* which develops into a new organism.

Sexual reproduction Reproduction usually involving two parents, which produce *sex cells*. These fuse together making a *zygote*, which develops through an *embryo* stage into a new organism.

Sieve tube A tube which forms part of *phloem* tissue. It transports food from regions of *photosynthesis* to growth and food storage regions.

Species A group of organisms which can mate together and produce young capable of further reproduction.

Sperms The male *sex cells* of animals.

Sphincter A ring of muscle found in the walls of tubular organs such as the digestive system. Its contraction slows or stops movement of substances through the tube.

Spiracle An opening on an insect's body through which air moves in and out of its *tracheal system*.

Spleen An organ immediately below the *stomach* which produces *white blood cells*, and destroys old, worn-out *red blood cells*.

Spongy mesophyll The layer of cells in a leaf immediately below the *palisade*. It contains large intercellular air spaces.

Spore A microscopic reproductive cell released from an organism during *asexual reproduction*. Typical of fungi, mosses, and ferns. In bacteria, a spore is a resting stage of the life cycle, usually formed when conditions are unfavourable for growth.

Stamens The male reproductive organs of a flower. The *anthers* of stamens produce *pollen* grains.

Stigma The part of a *carpel* to which *pollen* grains become attached during *pollination*.

Stimulus Anything which produces a *response* in an organism, e.g. a painful burn on the skin.

Stomach A bag-like organ at the end of the gullet.

Stomata Pores in the *epidermis* of plants through which air enters and leaves, and water evaporates during *transpiration*.

Suspensory ligaments Fibres which hold the lens in position within the eye.

Sweat gland A *gland* in the skin. It produces water which evaporates into the air and cools the body.

Swim bladder An air-filled bladder in the bodies of certain fish. It is used to adjust the density of the body until it equals that of water, making the fish weightless.

Symbiosis A close association between two different organisms in which both benefit.

Synapse A microscopic gap over which nerve impulses pass when moving from one nerve cell to the next.

Synovial joint Any freely moveable joint in the skeleton, e.g. the elbow.

Taste bud A collection of sensory nerve endings in the tongue which respond to certain chemicals in food, producing the sensation of taste.

Tendon A strong band of fibres which attaches muscles to bones.

Testis Male reproductive organ of animals. Produces *sex cells* called *sperms*.

Thorax (of insects) The middle segment of the body.

Thorax (of mammals) The cavity within the chest that contains the lungs, heart, and main blood vessels.

Thyroid An *endocrine gland* in the neck. It produces a *hormone* called thyroxin which has a major influence on physical and mental development.

Tissue A collection of similar cells which work together to perform a particular function, (e.g. muscle, bone).

Tissue fluid Fluid which is forced through *capillary* walls and moves between all *cells* of the body providing them with food and oxygen and removing their waste products.

Toxin A poisonous substance.

Trace elements Minerals which are essential for the healthy growth of plants but which are required only in minute quantities, (e.g. iron).

Trachea The windpipe.

Tracheal system A system of tubes through which air passes in and out of an insect's body.

Transpiration Evaporation of water from plant *cells* through *stomata*.

Tropism A movement in a plant in which the direction of root and shoot growth alters according to the direction of a *stimulus*.

Umbilical cord A tube containing blood vessels connecting a developing *embryo* with its *placenta*.

Ureter A tube which carries *urine* from a kidney to the bladder.

Urethra A tube which carries *urine* out of the body from the bladder.

Urine Liquid containing waste materials removed from the blood as it is filtered by the kidneys. It consists of water, urea, and various mineral salts.

Uterus A bag-like organ of the female reproductive system. It contains, protects, and nourishes the developing *embryo*.

Utricles Fluid-filled spaces in the inner ear containing sensory hair cells and *otoliths*. They detect acceleration and deceleration of the body and changes in its position relative to the pull of gravity.

Vaccine A suspension of dead, inactivated, or relatively harmless germs which, when injected into the bloodstream, stimulate the production of *antibodies* and make the body *immune* to attack from harmful disease organisms.

Vacuole Fluid-filled space in the *cytoplasm* of a *cell*.

Vascular bundle Strand of *xylem* and *phloem* tissues running from the roots into the leaves. It transports food and water throughout the plant and supports the softer tissues.

Vascular system (of animals) The heart and blood vessels.

Vector An animal which carries disease organisms.

Vegetative reproduction A form of *asexual reproduction* in which outgrowths from a plant eventually separate and continue an independent existence as new plants.

Vein (of animals) A vessel which carries blood towards the heart.

Vein (of plants) A strand of *xylem* and *phloem* tissue in a leaf.

Vena cava The main *vein* of the body.

Ventilation The movement of water past a respiratory surface such as a lung or gill, which enables *gaseous exchange* to take place.

Ventricle One of the large, thick-walled lower chambers of the heart that pump blood into *arteries*.

Vertebral column The backbone or spine. A chain of small bones called vertebrae that support the body, protect the spinal cord, and permit bending movements.

Villi Minute finger-like structures on the inner surface of the *duodenum* and *ileum*. These occur in millions, greatly increasing the surface area available for *absorption* of digested food into the blood.

Vitamins Chemicals required in small amounts to maintain health.

Vitreous humour The jelly-like substance that fills and supports the space behind the lens of the eye.

Weathering   The process by which rocks are broken down into small fragments, e.g. by wind, rain, and frost.

White blood cells   The general name for a number of different colourless *cells* in the blood, (e.g. *phagocytes* and lymphocytes).

White matter   Nervous tissue in the brain and spinal cord which consists of nerve fibres.

Xylem   A plant tissue which transports water and dissolved minerals from the soil to the leaves, and also supports the softer plant tissues. It consists of *xylem* vessels and fibres.

Zygote   The *cell* which results from the fusion of male and female *sex cells* (a fertilized egg).

Index

205